"Tim Lindsey's timely book is a gift from an active player in the field of positive sustainable growth. By sharing his wisdom and experience, he is showing us how to use principled values-based business practices to create profitable economic value."

**William McDonough, Chief Executive, McDonough Innovation, Yale University School of Architecture, co-author of *Cradle to Cradle* and *The Upcycle***

"Tim Lindsey makes a compelling case for business sustainability as a competitive advantage. He sees it as innovation, pure and simple, not just a philosophy or even a 'business case,' and articulates four principles that can be used as a guiding compass. Like other effective innovators, Lindsey finds opportunities in the constraints and describes how they can be turned into profitable strategies. This is a great read for sustainability leaders, or anyone who wants to become one."

**Joel Makower, Chairman and Executive Editor, GreenBiz Group, and co-author of *The New Grand Strategy: Restoring America's Prosperity, Security and Sustainability in the 21st Century* (St. Martin's Press)**

"Dr. Lindsey makes a thoughtful and compelling case that the sustainability of business practices is critical for the success of organizations in the coming decades."

**Jeff Brawn, Professor and Head, Department of Natural Resources and Environmental Sciences, University of Illinois at Urbana-Champaign, USA**

"A masterful pragmatic hands-on approach to advancing sustainability, *Headwinds of Opportunity* is a must-read for all engaged in the essential effort to achieve a sustainable future. This is not just for the world of commerce but for all concerned with a sustainable future for all."

**Thomas E. Lovejoy, University Professor of Environmental Science and Policy, George Mason University, and 2012 Blue Planet Prize Laureate**

"*Headwinds of Opportunity* makes a compelling case for business sustainability as a competitive advantage. Tim Lindsey not only provides the strategic why, but also delivers on the practical how-to for making sustainability the innovation of our time. I highly recommend this book to management everywhere looking at how to better position their business to thrive and truly have an impact."

**Todd Thornburg, Energy Efficiency Portfolio Manager, ComEd, USA**

"When I think of headwinds, I think about an unseen force that impedes progress. Dr. Lindsey's book perfectly describes these forces, articulates the impacts they will have on business of all types, and offers a clear path forward. Like the Hippocratic Oath of first doing no harm, this book lays out an oath all companies should take: Prevent waste, improve quality, optimize systems and restore value."

**Rich Goode, Executive Director, Climate Change and Sustainability Services, Ernst & Young, USA**

"Tim Lindsey produces an anthology of exceptional sustainable innovation case studies and global market practices that can serve as an excellent handbook to anyone interested in applying sustainable innovation in business strategy. *Headwinds of Opportunity* touches upon the most crucial sustainability challenges of our times showing the way to competing on growth."

**Michael Spanos, Managing Director, Global Sustain Group**

"Masterfully done, Tim Lindsey's work now shows all of us not only how, but why the business community should chose this path. Tim takes the initiative and offers the case study proofs that double down on both of the most important business concepts for success – how to create lasting value and to do that innovatively. As you read his work, you will likely unlock the power of 'why' for your business, and adopt a philosophy of 'why wouldn't we?'"

**Dr John Mogge, Environmental Market Director, CH2M, USA**

"Tim Lindsey's guide to integrating sustainability into core business strategy is straightforward, insightful, and pragmatic. He demonstrates a keen understanding of aspects of current corporate culture that impede opportunities and he lays down a challenge for a bold, disruptive, and innovative approach to turn headwinds into tailwinds. A valuable read for anyone who is truly interested in driving sustainability improvements within their organizations."

**Kevin McKnight, Vice President, Environment, Health & Safety and Chief Sustainability Officer at Alcoa (Retired), USA**

"Tim Lindsey makes a compelling business case for why tomorrow's successful companies and organizations will be those that proactively adapt themselves to the changing winds that are shaping global conditions, opportunities, and markets. More than this, he provides important, practical insights and examples for how those organizations will succeed in integrating sustainability principles and practices within the fabric and sails of their enterprise."

**Todd S. Bridges, Senior Research Scientist, US Army Corps of Engineers, USA**

"*Headwinds of Opportunity* presents a clear and compelling case for implementing sustainable strategies as the centerpiece of our business organizations. Thoroughly researched and brilliantly crafted, this book will awaken a new paradigm of thinking – and doing – for the 21st century. Thank you, Tim, for the gift of these powerful, pointed lessons."

**Timothy Hoerr, CEO and Managing Partner, Serra Capital, USA**

"Tim Lindsey offers a new insight and an impartial approach for suppliers and customers to work together to review all aspects of the supply chain in order to provide a sustainable approach to delivering exceptional value. A great read for all functions within an organization seeking to drive a competitive advantage."

**Neil Winterbottom, Director of Operations, Fluidcare Americas Houghton International Inc., USA**

"*Headwinds of Opportunity* is a humane, understandable, and important book by a major practitioner of corporate practice. Tim Lindsey's book, based on 35 years of astute observation of the turbulent worlds of corporate behavior and applied research, teaches firms how to improve competitiveness by addressing the challenges that face us all regarding energy, minerals, water and food scarcity, without jargon or pretense. The book delivers by describing the facts and moments in recent history when new significant corporate opportunities arise due to mounting changes in public expectations. It is in this flux of public expectations that both the destiny of great firms and the capabilities of individuals flourish as they are shaped into our shared future."

**Bruce Piasecki, Founder of the AHC Group's Corporate Affiliates workshops, and author of *Doing More with Less: The New Way to Wealth***

"In this book, Tim Lindsey brings his experience of innovation and sustainability from the field, from industry and academia, and finally from leading the implementation of sustainability practices and culture at Caterpillar. Like many innovators, he sees opportunity in the threat: in his analysis the headwinds that are challenging industrial progress turn into headwinds of opportunity. This book will be of great value to anyone who is engaged in making the shift to sustainability – or for that matter, in deep organizational changes of any sort."

**Beebe Nelson, co-author of *New Product Development for Dummies* (2007, Wiley) and *Innovation Governance* (2014, Wiley)**

"Tim Lindsey captures the concrete business case for sustainability through compelling and tangible case studies and examples. A must-read for anyone in the private or public sector needing to make the case for sustainable solutions."

**Ed Pinero, CEO, The Pinero Group LLC and former Senior Vice President for Sustainability and Public Affairs, Veolia, North America**

"In *Headwinds of Opportunity*, Tim Lindsey crystallizes the difference between doing things right and doing the right things, framing sustainability, why it's imperative, and what it requires of us. With deep experience and a sharp eye, he offers readers a compass to the thinking and practices that teach us how to score in addition to keeping score. In a nutshell, he describes how value can and must be both created and fed back in a circle for prosperity that actually lasts."

**Todd Fein, CEO, Green Diamond and former US Department of State Franklin Fellow for Global Sustainability Reporting**

"Tim Lindsey's *Headwinds of Opportunity* is not only a crucial read in applying sustainable innovation in business strategies, but it is a gift, depicting an elevated form of attentiveness and awareness towards our environment and what our true responsibilities are within the systems in which we live and work. Having worked for the US Army Corps of Engineers, an agency responsible for both our water infrastructure and water resources and now returning to the private sector, I can truly say that Tim's assertion that infrastructure can act as a physical nexus or a primary interface between the economic, social, environmental, and business assets, is spot on. The sustainable intelligence written in *Headwinds* can serve to break down the walls between the private sector's agility and federal agencies struggling with massive aging assets and an uncertain water future."

**Heather M. Morgan, Sustainability Manager, AECOM, USA**

# Headwinds of Opportunity

A Compass for Sustainable Innovation

# Headwinds of Opportunity

## A Compass for Sustainable Innovation

TIM LINDSEY

LONDON AND NEW YORK

First published 2018
by Routledge
2 Park Square, Milton Park, Abingdon, Oxon OX14 4RN

and by Routledge
711 Third Avenue, New York, NY 10017

*Routledge is an imprint of the Taylor & Francis Group, an informa business*



*British Library Cataloguing-in-Publication Data*
A catalogue record for this book is available from the British Library

*Library of Congress Cataloging-in-Publication Data*
A catalog record for this book has been requested

ISBN: 978-1-78353-806-5 (hbk)
ISBN: 978-1-78353-760-0 (pbk)
ISBN: 978-1-78353-761-7 (ebk)

Typeset in Sabon and Caecilia
by BBR Design, Sheffield

# Contents

# Figures

# Tables

# Preface

Since starting my career in 1980, I have served in a variety of roles in the private sector and academia that afforded me the opportunity to study, analyze, and drive change regarding various aspects of business resourcefulness and innovation. These experiences have run the gamut in terms of extremes. I started my career working for an engineering/construction firm that was tasked with cleaning up "Superfund" hazardous waste sites. This experience forced me to deal with the impacts that extreme levels of human abuse and neglect can have on communities and the environment. Working to restore God's creation, while wearing protective gear that one would normally expect to see on astronauts, left a lasting impression on me. On one occasion, the protection afforded by the sophisticated equipment proved to be inadequate for me and some of my colleagues. Toxic fumes from a ruptured container penetrated our face masks and respirators and we were forced to inhale our own vomit as we frantically dealt with an unexpected surge of extremely pungent emissions. That day, in particular, stands out as a moment of clarity, when I realized there had to be a better way to deal with these types of problems.

I left the exciting and dysfunctional world of Superfund site clean-ups in 1984, in search of less hazardous work. I relocated to the Powder River basin of Wyoming where I spent seven years working for the world's largest energy company at a coal mine. I served in various environmental, safety, and production-focused roles. These assignments afforded me the opportunity to directly participate in meeting the day-to-day production and quality demands of a complex business operation while protecting the health and safety of employees, and preserving and restoring environmental resources.

My journey continued to the University of Illinois where, for over 20 years, I had the opportunity to conduct applied research and provide outreach to diverse business interests regarding the methods and innovations needed to improve their sustainability performance. This experience allowed me the opportunity to study in considerable detail the interrelationships between science, engineering, economics, nature, and society. I was able to assimilate what I learned, and to develop and assess various strategies and tactics for conducting business across multiple sectors in ways that are beneficial to the bottom line, the top line, communities, and the environment. I used the lessons learned from this experience to cultivate and test a set of core principles for driving innovation in ways that can improve long-term effectiveness and competitiveness for virtually all organizations.

In 2012, I left my academic career and returned to the private sector where, for four years, I served as the Global Director of Sustainable Development at Caterpillar Inc., a Fortune 60 company. At Caterpillar, I had the opportunity to take the principles of innovation and business sustainability I had honed over my career and work with key leaders to embed them into the corporate culture. This approach produced dramatic changes at Caterpillar. Within two years, the corporation officially recognized sustainability as the fifth core value for the enterprise and incorporated it into the employee code of conduct, goal-setting process, and performance review process. We developed consistent messaging regarding the meaning and importance of sustainability to the enterprise. We prepared a training module and employee engagement program to help employees embed sustainability principles into their daily work.

We were able to implement sustainability principles in ways that would drive innovation and improve competitiveness throughout the corporate value chain. This included incorporating sustainability into corporate roles (research, product development, human resources, accounting, strategy, and risk) and operational functions (supply chain, manufacturing, quality, logistics, dealerships, and customer processes). We developed a portfolio of sustainable products and services that accounted for 18% of total corporate revenue. During my tenure there, Caterpillar achieved elite recognition for its sustainability achievements, including:

- Top 4% ranking for our sector on the Dow Jones Sustainability Index (after being removed from the index a few months after I arrived)
- Runner-up status for circular economy achievements ("The Circulars") at the 2016 World Economic Forum
- Keep America Beautiful's 2015 "Vision for America" award

My journey has led me to work with hundreds of organizations, across multiple sectors, to help them improve their resourcefulness. I have actively participated in a great deal of history during society's ongoing transition toward a more sustainable culture and I have witnessed many important successes and failures. I have learned a great deal from both and have developed a strong sense of what approaches are most effective for various types of organization, in both the short and long term.

Society faces escalating challenges from the effects of increasing population and declining resources. My journey has led me to observe a number of incontrovertible truths, which can be applied broadly to break down challenges into manageable components and develop effective strategies for addressing them. Organizations commonly exhibit four behaviors that hinder their effectiveness and create widespread constraints to economic growth, community development, and environmental protection. To address these behaviors, I have documented a set of four guiding principles that organizations can follow to correct their behaviors, improve their effectiveness, and enable widespread sustainable progress.

I have written this book in an effort to share the important lessons, strategies, and principles I learned over my more than 35-year career. My hope is that readers can apply this knowledge in practical ways that simultaneously improve their long-term sustainability performance and overall effectiveness. Great care was taken to avoid jargon and keep the language understandable and compatible with concepts that are commonly used in business. The book is structured to provide a compass that leaders can use to:

1. Identify and assess emerging trends, constraints, and innovations that can affect organization performance
2. Develop strategies based on sound principles that address the growing constraints and take advantage of emerging opportunities
3. Use proven sustainability principles to drive innovation and improve effectiveness

Real-world examples of how forward-thinking organizations have successfully put these concepts into practice are provided along the way. Developing and following strategies based on the guidance provided in this compass will help leaders produce more enduring performance gains that improve competitiveness throughout their value chain.

# 1

# Increasing headwinds

> The wind blows where it wishes and you hear the sound of it, but do not know where it comes from and where it is going.
>
> John 3:8

Business leaders and members of the media that report and forecast business performance regularly talk in terms of "headwinds" whenever significant challenges emerge. Here are a few examples, clipped from recent headlines, that show how the headwinds metaphor is commonly used to characterize emerging challenges:

> "The economy still faces major headwinds"
>
> "We are struggling against headwinds from higher geopolitical risk"
>
> "We're facing currency headwinds"
>
> "We continue to face headwinds from uncertainty"
>
> "Numerous headwinds on the horizon"

> "We face stiff headwinds from rising raw material and energy costs"
>
> "We see regulatory headwinds coming"
>
> "They face headwinds from mostly brand and distribution problems"
>
> "Too many headwinds to ignore"
>
> "The stock market faces considerable headwinds ahead"
>
> "This legislation faces political headwinds"
>
> "They continue to face significant headwinds from the employment front"

Typically, the term "headwinds" is used as a metaphor to describe factors that are perceived to be beyond our control. Just as we can't control the weather, describing factors as "headwinds" implies that the only way to manage them is to either (staying with the winds metaphor) adjust sails and change course, or hunker down and batten down the hatches until the storm passes and conditions improve. Examples of factors that have commonly been described as headwinds include unfavorable conditions associated with the economy, currency value, cost of commodities, new regulations, new legislation, uncertainty, or a downturn in a key market. Because of their magnitude, it is often difficult for individual people or organizations to take actions that have a significant impact on these conventional sources of headwinds.

## A special type of headwind

A special type of headwind has emerged in the past few decades that is not going away. These headwinds can be relentless and will punish organizations that try to tackle them head on. They are a product of unfavorable trends and interactions across a broad combination of social, environmental, and economic challenges. I refer to this new category as **"constraint headwinds"** because they are constraining the health and productivity of the economy, the environment, communities, and citizens in many parts of the world. Constraint headwinds are produced by two converging trends:

1. An increase in population and demand for resources
2. A decline in the quality and availability of resources

As these trends converge, they trigger a variety of outcomes and issues that negatively impact the economy, the environment, communities, and citizens. As shown in Figure 1.1, these outcomes and

FIGURE 1.1 **Cascading issues that generate constraint headwinds**

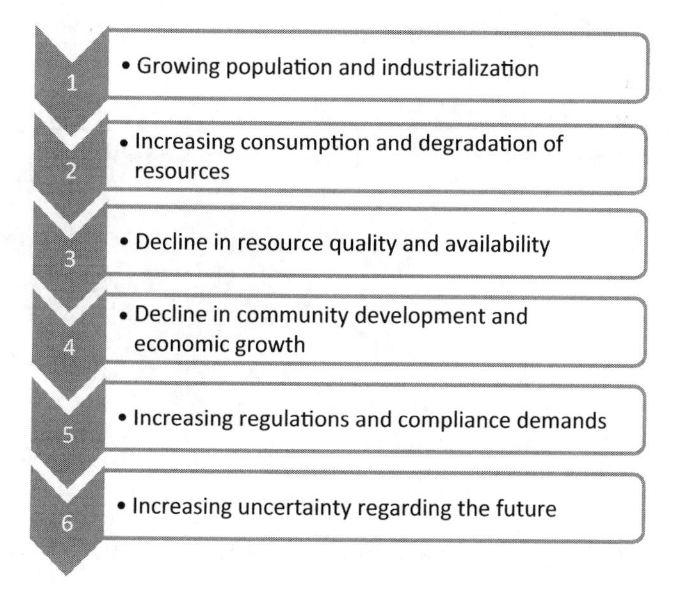

issues cascade from one to another in a chain reaction that produces a diverse mix of constraints that, in turn, create formidable headwinds for the economy and degrade the quality of communities and the environment.

Constraint headwinds have grown in magnitude and intensity during the past several decades and they are frequently a primary contributor to the aforementioned conventional headwinds because they can affect everything from regulatory policies to the cost of commodities. Figure 1.2 shows how constraint headwinds emerge and grow over time.

Businesses, in particular, contribute to the magnitude and intensity of constraint headwinds through management choices regarding their processes, products, and systems. These actions and outcomes can generally be lumped into four broad categories as follows:

FIGURE 1.2 **Sources of constraint headwinds**

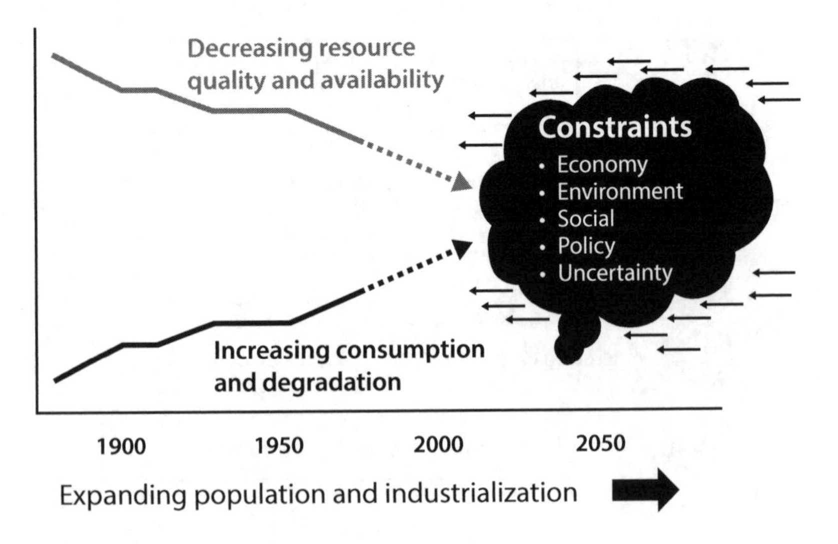

1. Wasteful practices
2. Deficient processes and products
3. Dysfunctional interactions
4. Negligent decisions

Management choices that lead to these actions and outcomes contribute directly to constraint headwinds and are described in more detail in Chapter 6. Many businesses and, in some cases, entire sectors (e.g., coal mining) are now facing constraint headwinds that jeopardize their future existence altogether because their management choices have increasingly led to these undesirable actions and outcomes.

While some organizations have been paralyzed by increasing constraint headwinds, others are thriving because they have been able to respond with measures that actually improve their effectiveness and competitive advantage. Forward-thinking organizations have been able to modify their processes, products, and systems in ways that diminish constraint headwinds and create prosperous opportunities for improving effectiveness.

As constraint headwinds have grown, there has been a corresponding increase in innovation focused on ideas, methods, materials, machines, and devices that address the problematic constraints at their root causes. In some cases, organizations have shifted their business models so fundamentally that the constraint headwinds that previously hindered their progress have now become prosperous sources of tailwinds that contribute to their success. Accomplishing such a dramatic shift requires a thorough understanding of the evolution of constraint headwinds and the factors that have contributed to their escalation. These aspects are described in the next two chapters.

# 2

# The evolution of constraint headwinds

> Nothing is better than the wind to your back, the sun in front of you, and your friends beside you.
>
> Unknown

In the early stages of industrialization, innovation was focused primarily on improving the productivity of personnel and capital assets. Lack of access to capable workers often hindered industrial growth, particularly in the New World. This constraint drove the development of diverse industrial innovations such as machinery, materials, and methods that could maximize workforce productivity. The natural resources needed to drive industrialization (energy, metals, lumber, water, etc.) were much more abundant than personnel, so innovators tended to focus less attention on the productivity and efficiency of these resources.

## Expanding industrialization and impacts

Industrial innovations required considerable inputs of concentrated energy to drive their productivity. Since labor access was limited, businesses needed to find alternative sources of energy to increase their productivity. Therefore, entrepreneurs pursued innovations that increased their access to reliable forms of concentrated energy that could be harvested, processed, and delivered in the quantities needed to expand production. Energy was harvested from an evolving mix of resources starting with flowing rivers that turned simple mill wheels, then transitioning to coal, oil, natural gas, and nuclear fuels. The increasing access to diverse energy sources enabled industries to produce the massive quantities of heat, steam, and electricity they needed with increasing intensity and reliability. Electronic innovations emerged along the way, and were later combined with information technology to automate production and create additional productivity gains. The combined effects of these innovations led to production breakthroughs in diverse sectors, including transportation, manufacturing, agriculture, forestry, and mining. For example, from 1961 to 2009, food production increased by 150% while the world's cropland grew by just 12% (FAO, 2011, pp. 1-59).

Supporting rapid growth in these diverse sectors required expanding access to raw materials, energy, processing capabilities, logistics, and customers. The mining and timber industries grew dramatically to feed the demand for raw materials and energy while lands were cleared and developed to expand agricultural production. The rapidly expanding flow of raw materials and products increased the need to construct various forms of infrastructure to support transportation, energy transmission, communications, water supply, and sanitation. Businesses and communities could not usually develop the much-needed infrastructure on their own. Consequently, they engaged with governments at various levels to assimilate and administer the resources required for infrastructure development and construction.

Infrastructure done well can be a great enabler that improves the quality of life in communities by increasing access to energy, transportation, goods, services, water, sanitation, education, and employment. Well-developed infrastructure has proven to be critical for supporting the growth of commerce by providing the connectivity needed for suppliers to deliver the raw materials that industry needs for production. Infrastructure also enables industries to distribute their products efficiently to an expanding customer base.

The 20th century was a time of remarkable progress for human civilization. The growing production and consumption of goods and services, combined with rapid growth in the construction of infrastructure and communities, led to expanding economies, employment opportunities, wealth generation, and an expanding tax base. An affluent middle class emerged and many millions of people benefited from the unprecedented economic growth and improved quality of life that resulted from expanding industrialization and infrastructure. However, in addition to the many benefits provided by increasing industrialization and infrastructure development, a variety of unanticipated side-effects emerged.

## Unintended consequences grow

> He who troubles his own house will inherit wind,
> and the foolish will be servant to the wisehearted.
> Proverbs 11:29

Early industrial systems tended to be wasteful because they functioned through an inefficient and unbalanced linear flow of inputs and outputs. Raw materials entered the systems while products and wastes exited. Products were shipped to customers while wastes were buried, burned, and/or discharged to the environment. In the relatively few instances where recycling took place, it was accomplished predominantly by entrepreneurs that functioned outside the industrial system. For instance, a producer of rubber floor mats might collect worn tires from an automotive repair shop to use as raw material whereas scrap rubber from actual tire production was sent to landfills.

Producers used increasingly sophisticated processes to convert inputs of materials, chemicals, and energy into products to meet a surging consumer demand. The complexity and dangers of industrial processes expanded rapidly, leading to increased frequency of accidents and injuries. By-products were commonly generated for which businesses could not find markets, so they accumulated as wastes in volumes that required disposal by whatever means were available, including incineration, burial in landfills, and discharge into rivers, lakes, and seas. Emissions from evaporation, combustion, and other reactions escaped into the atmosphere. At the end of the products' useful life, consumers usually disposed of the once valuable items as garbage, thereby removing the raw materials from the value chain and rendering them unavailable for use in future production.

The impacts from population growth, industrialization, and infrastructure development have varied from region to region due to multiple factors. Demographics and economic development drive demand while advances in technology affect costs and impacts. The levels of prosperity that were created from industrialization and infrastructure development have similarly varied greatly from region to region and some areas pretty much missed out on the benefits altogether. Regions with ready access to the critical inputs needed for industrial growth (energy, iron/steel, water, productive land, etc.) benefited considerably more than regions that lacked these resources. This disparity has historically been one of the primary factors that has created separation between the developed and developing worlds with respect to unequal levels of prosperity and quality of life.

Early efforts to develop infrastructure systems did not fully consider all the potential impacts and consequences that could negatively affect communities and the environment. **Infrastructure frequently provides the primary interface between the economic, social, and environmental dimensions of society.** The methods used to design, construct, operate, and maintain the infrastructure have proven to be critical for ensuring the quality of all three dimensions. Many early infrastructure projects did not incorporate sound development and management practices for communities, lands, water bodies, and air resources. As a result, **infrastructure projects often created a variety of dysfunctional interactions between community, environmental, and business assets.**

Failures associated with early infrastructure development often contributed unnecessarily to urban sprawl, congestion, flooding, air and water pollution, ecosystem destruction, and losses of biodiversity. For example, the USEPA reported in 2016 that 41 states reported higher than acceptable levels of lead in drinking water from 2012 to 2015 as a result of corroding pipes and plumbing that were installed decades earlier (Gusovsky, 2016). Infrastructure construction has historically resulted in considerable consumption and degradation of resources. The United Nations (UN) estimates that, from 1900 to 2005, annual extraction of construction materials has grown by a factor of 34, ores and minerals by a factor of 27, fossil fuels by a factor of 12, biomass by a factor of 3.6, and total material extraction by a factor of about 8, while gross domestic product (GDP) rose 23-fold (UNEP, 2011, pp. 55-69).

As long as resources were plentiful and population was sparse, the wasteful side-effects and impacts of industrialization, such as industrial accidents, land degradation, resource depletion, waste generation, emissions, and pollution of air and water, were usually tolerated. They were generally considered to be part of the cost of progress and the ends were largely perceived to justify the means. However, as population and industrialization continued to increase, so did their impacts and consequences. Advances in industrial productivity generated an increasingly diverse and complex mix of hazards, side-effects and by-products that led to unintended consequences for communities, citizens, and the environment. Innovations in chemicals, materials, processes, machines, and energy continuously generated new challenges.

The unintended consequences of industrialization often outpaced society's ability to manage them. A considerable knowledge gap developed between the technical complexity and risks associated with the new issues and society's ability to mitigate them. New measures were continuously needed to address the various hazards, wastes, and emissions associated with the dysfunctional industrial systems. Governments attempted to intervene by establishing and implementing policies, standards, regulations, and controls that could ensure the safety of people, communities, and the environment. Unfortunately, their efforts often lagged behind because technical solutions were not available or they were perceived to be complex and costly. As a result, resource availability and quality diminished dramatically in some regions, along with the quality of the environment and the quality of life in affected communities.

## Escalating impacts become unacceptable

By the latter half of the 20th century, the complex mix of hazards, wastes, and emissions associated with dysfunctional industrial systems had grown to levels that society could no longer ignore. The U.S. Occupational Safety and Health Administration (OSHA) estimates that by 1970, about 14,000 workers were killed on the job each year in the U.S.A. (OSHA, 2017). Work-related diseases are more difficult to estimate than accidents but the International Labour Organization (ILO) estimates that they kill six times more people each year than accidents. In total, about 2.34 million people die each year (about 6,400 per day) from work-related accidents and diseases (ILO, 2013). A variety of occupational lung diseases became prevalent, including:

- **Asbestosis**: Asbestos miners and those who work with asbestos insulation
- **Pneumoconiosis** (black lung disease): Coal miners
- **Silicosis:** Quarry and tunnel operators
- **Byssinosis:** Workers in the cotton textile industry

Additionally, some of the thousands of new chemicals that entered markets through industrialization have been linked to cancer. On-the-job accidents, injuries, and illnesses create considerable hardship for victims, their families, communities, and employers. Not only do the health and safety issues have devastating social impacts, they also add considerable monetary costs. Globally, about 4% of global GDP (about US$2.8 trillion) is lost annually as a result of occupational accidents (ISSA, 2017). According to the U.S. Consumer Product Safety Commission, deaths, injuries, and property damage from consumer product incidents cost the nation more than US$1 trillion annually (CPSC, 2017).

By the 1980s, U.S. public awareness of the growing impacts from wasteful and negligent practices had reached a pinnacle. A rapidly growing number of toxic waste sites exhibited alarming levels of toxins from years of illegal waste dumping. In some regions, rivers became so polluted that fishing had to be prohibited. Some contaminated sites proved to be particularly problematic and garnered considerable media attention. Examples include:

- **Cuyahoga River fire:** In 1969, the Cuyahoga River actually caught fire in Cleveland when floating pieces of oil-slicked debris were ignited and flames reached heights of over five stories (Ohio History Central, 2017).
- **Love Canal:** In 1978, at the Love Canal site near Niagara Falls New York, the USEPA evacuated over 200 families and a public school that were located near an illegal industrial dump that had been used for decades by the Hooker Chemical Company. The wastes contained 82 different compounds, 11 of them suspected carcinogens which were leaching into the basements of many residents (Beck, 1979).
- **Valley of the Drums:** In 1979, a 13-acre valley in Bullitt County, Missouri that had been filled with over 17,000 drums of toxic industrial waste had to be remediated after receiving national attention (Bullitt County History, 2017).
- **Times Beach:** In 1983, the 2,000 residents of Times Beach, Missouri, located 17 miles southwest of St. Louis, had to be permanently evacuated due to a dioxin contamination associated with the illegal spraying of contaminated waste oil on dirt roads (EPA, 1983).

Globally, a series of disasters occurred in the 1980s that raised public awareness of the growing problems and contributed to the growing constraint headwinds (World Commission on Environment and Development, 1987, pp. 1-6). Examples include:

- A leak from a pesticide factory in Bhopal, India, killed more than 2,000 people and blinded and injured over 200,000 more
- Liquid gas tanks exploded in Mexico City, killing 1,000 and leaving thousands more homeless
- The Chernobyl nuclear reactor explosion in the Ukraine killed 31 people and sent nuclear fallout across Europe
- During a warehouse fire in Switzerland, agricultural chemicals, solvents, and mercury flowed into the River Rhine, killing millions of fish and threatening drinking water in the Federal Republic of Germany and the Netherlands
- An estimated 60 million people died from diarrheal diseases in the developing world as a result of unsafe drinking water and malnutrition; most of the victims were children

The growing frequency and intensity of these types of incidents and their relentless side-effects and impacts are at the core of constraint headwinds. Not only did the impacted regions experience negative safety, health, and environmental effects, they also experienced diminished economic growth, employment opportunities, and tax revenue. As a result, society became increasingly aware of the interrelationships between environmental, social, and economic aspects and their interdependence on each other.

## Society responds to increasing constraints

> A great wind is blowing, and that gives you either imagination or a headache.
>
> Catherine the Great

As the impacts from dysfunctional industrial systems increased in both frequency and intensity, various coalitions surfaced to advocate for better protection of communities, citizens, and environmental resources. They pushed for change through a combination of policy and economic reforms, and policy-makers responded by developing various regulatory measures to address the issues. However, in spite of the obvious need for changes, the proposed policies and regulations were often met with considerable resistance because the issues were complex and represented by diverse stakeholders. New regulations were frequently perceived to be burdensome and ineffective by the regulated groups. Consequently, consensus on a path forward was often difficult to achieve.

Attempts by governments to address the growing constraints with policies and regulations achieved mixed success. Due to the hazardous nature of many of the problems at hand, the initial regulatory focus was directed primarily at implementing measures that protect people. The regulations and enforcement were fairly effective with respect to halting criminal and negligent behavior and responding to urgent and dangerous issues. However, most regulations responded to symptoms of deeper issues and failed to address the problems at their roots. They often added considerable complexity and uncertainty to markets and business processes.

In response to dangerous working conditions across the U.S.A., and as a culmination of decades of reform, the Occupational Safety and Health Act of 1970 was signed into law by President Nixon.

This law led to the establishment of OSHA, the National Institute of Occupational Safety and Health, and the independent Occupational Safety and Health Review Commission. OSHA was established to ensure safe and healthful working conditions for working men and women by setting and enforcing standards and by providing training, outreach, education, and assistance. Since OSHA's establishment, workplace fatality rates and occupational injury and illness rates have declined by about two-thirds while U.S. employment almost doubled (OSHA, 2001).

Despite OSHA's significant success, businesses have regularly criticized it as an example of excessive government regulation. OSHA standards are often perceived to be expensive to implement and some of the measures required for compliance are not effective. Careful studies of the impact of inspections on large employers in sectors where OSHA has concentrated its efforts, such as some specific manufacturing and construction industries, indicate that repeated OSHA inspections at the same job site have little impact on changing either compliance with standards or injury and illness rates (Gray and Mendeloff, 2005). Industry safety experts regularly argue that OSHA's narrow focus on improving the safety of working conditions limits its effectiveness because most workplace accidents occur as a result of unsafe behaviors by workers.

The growing need to address the magnitude and severity of expanding waste and pollution problems led to the creation of agencies such as USEPA (established in 1970 by the Nixon administration). Initially, regulators focused on protecting human health and the environment by addressing the most severe problems at hand. Their first priorities were to prosecute environmental criminals and clean up severely contaminated lands and waters. In 1980, USEPA was empowered to systematically address the growing need to remediate the large number of contaminated sites through implementation of the Comprehensive Environmental Response, Compensation, and Liability Act of 1980 (CERCLA). Commonly referred to as "Superfund," CERCLA enables the U.S. Environmental Protection Agency (EPA) to clean up such sites and to compel responsible parties to perform clean-ups or reimburse the government for EPA-led clean-ups (EPA, 2016).

In addition to prosecuting environmental criminals and cleaning up contaminated sites, policy-makers also proceeded with developing a system of regulations and controls focused on addressing the growing quantity and complexity of wastes and emissions. The command and control tactics they developed provided an "emergency response" that slowed the rapidly escalating problems. They initially established methods for treating and managing the waste, emissions, and pollution, then worked their way upstream toward the actual sources of the problems: the processes, products, and systems associated with industrialization.

The expanding emphasis on regulatory requirements and controls generated some unintended consequences that further increased complexity and constraints. Many of the problems facing society were already very technically complex and difficult to solve. Adding an onerous regulatory dimension to the mix of issues created, in effect, another layer of complexity for regulated organizations to deal with. Additionally, the command/control approach did not promote cooperation between the regulators and the organizations they regulated, and led to polarization of the various stakeholders. The net effects of the expanding regulatory system further intensified the growing constraint headwinds, leading some businesses to relocate to the developing world in order to avoid the additional constraints.

Government agencies generally lack deep expertise regarding the sources of the problems they are tasked with regulating – the industrial processes, products, and systems that generate accidents, wastes, and emissions. Consequently, regulators tend to focus their attention on developing general prescriptive requirements that have broad applicability. The requirements are targeted at treating, controlling, and managing issues from the perspective of the people, communities, and environment that regulators are charged with protecting. Frequently, the regulatory requirements have created compliance challenges for business because their prescriptive nature often doesn't allow for flexibility that could address the issues more proactively and efficiently. As a result, compliance with early safety and environmental regulations tended to be inefficient, expensive, and not fully effective.

In the case of hazardous waste regulations, the prescribed measures often led to simply transferring pollution from one medium to another. For instance, burying liquid wastes contained in new metal drums – in well-constructed landfills – kept the wastes out of water bodies in the short term. But the drums eventually rusted and the poisons leached through the landfill liner and into groundwater. As the costs of managing emissions and wastes escalated, the private sector has increasingly objected to the requirements' impacts on their operations and their bottom line. Some business groups have attempted to address the additional cost burden by lobbying for less stringent regulation and enforcement, often claiming that they can't compete with foreign businesses that do not have to incur the added costs.

## Complex problems produce conflicts and challenges

Conflicts have been common between environmentalists and business leaders. Conventional wisdom on both sides usually assumes that economic development and environmental protection are at

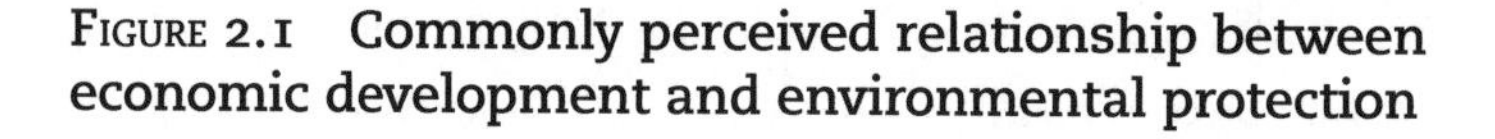

FIGURE 2.1 **Commonly perceived relationship between economic development and environmental protection**

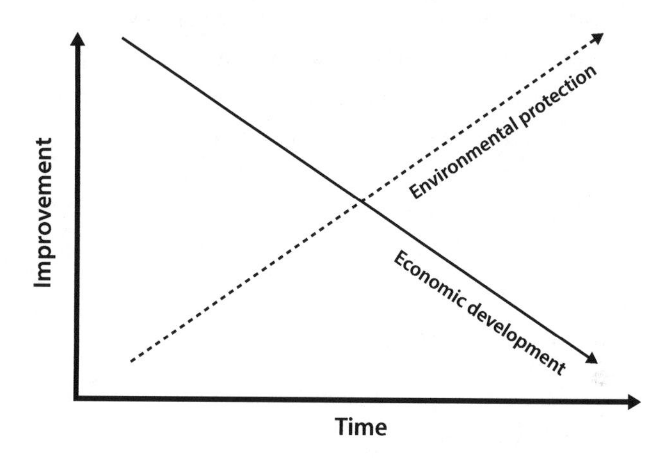


odds with each other. This assumption is based on decades of experience and dozens of examples where policies focused on improving one dimension resulted in negative impacts to the other. Figure 2.1 shows how the relationship between economic development and environmental protection has historically been perceived.

Policy-makers have continuously struggled to balance the needs for economic growth with their efforts to protect the environment and the people who depend on it. This approach has historically been contentious as the vast majority of decisions produced winners and losers.

The trend toward increasing constraint headwinds will likely intensify over the next several decades. Global population continues to grow and access to needed resources is challenged by increasing consumption and degradation. New issues and impacts regularly surface from today's complex global economy, and governments continually respond to the new challenges with policies and regulations that grow in volume, complexity, and compliance burden.

Businesses struggle to respond to the growing needs because of uncertainty regarding rapidly changing markets, growing regulations, and shareholder demands. The diverse collection of environmental, social, and economic impacts that aggregate to form constraint headwinds is an increasing concern for governments, communities, and businesses all over the world.

# 3

# Current and future constraint headwinds

> Human misery must somewhere have a stop; there is no wind that always blows a storm.
>
> Euripides

## Intensifying social and environmental constraints

The global population reached 7.3 billion as of mid-2015 after adding about 1 billion people since 2003 and 2 billion since 1990. Population growth is projected to continue for many years and reach 9.7 billion in 2050 and 11.2 billion in 2100 (UN Department of Economic and Social Affairs, Population Division, 2015). In addition to the growing population, consumption patterns in many regions have increased even more rapidly, particularly in Asia. The global middle class expanded from 21% of the total population in 1990 to 56% in 2008 (Amoranto *et al.*, 2010). Simultaneously, there has been a significant decrease in the share of people living in extreme poverty, declining from 55% to 24% during the same period (Asian Development Bank, 2010). These trends are expected to continue in the future with the middle class expanding from just over 2 billion in 2010 to 4.9 billion by 2030 (Kharas, 2010, pp. 17-29). Not only will population growth continue, it will increasingly shift from less-populated rural settings to more densely populated urban environments. Population in urban areas is projected to increase by 2.5 billion people by 2050, with nearly 90% of the increase concentrated in Asia and Africa. Megacities (cities of more than 10 million people) are also on the rise. In 1970, the world had only two megacities – Tokyo and New York City – but today there are 28 and this number is increasing. By 2030, the world is projected to have 41 megacities (UN Department of Economic and Social Affairs, Population Division, 2015).

In many circumstances, the infrastructure of these cities is unable to keep pace with such rapid increases in population – nor the needs of their inhabitants. This has implications for the quality of life of city dwellers and their access to basic amenities. Over 1 billion people already live in city slums (UN Population Fund, 2007, pp. 15-16) and (unless massive investments are made in infrastructure and services) their numbers will greatly increase along with the proliferation of social problems. Innovations in materials, chemicals, and processes continue to create complex new products with characteristics that raise health and safety concerns (e.g., foods that contain genetically modified organisms). Additionally, new sources of wastes and emissions (e.g., ozone-depleting solvents) continuously create new challenges. Technology shifts associated with rapidly changing products and the processes used for their production have also resulted in new challenges (e.g., the rapid growth of electronic waste).

Continued population growth and consumption are projected to intensify resource constraints and produce a variety of new risks and uncertainties regarding international relations in an already turbulent world. Water and land are already scarce in many parts of the world and are coming under pressure from competing users as urbanization and industrial development continue. One person in ten lacks access to sanitary water and one in three lacks access to a sanitary toilet (WHO and UNICEF, 2015, pp. 6-12).

The latter half of the 20th and the early part of the 21st century have wreaked havoc on global wildlife diversity. In less than two human generations, populations of many vertebrate species dropped by half. Population declines averaged 58% from a sampling of more than 10,000 mammals, birds, reptiles, amphibians, and fish from 1970 and 2012. Habitat loss and degradation, and exploitation through excessive hunting and fishing, are the primary causes of these declines (WWF, 2016, pp. 58-87).

Centuries of negligent land management practices have led to massive degradation on all continents. Over 25% of the Earth's land (an area approximately the size of the North American continent) is now considered to be "highly degraded" due to poor management practices that have resulted in deforestation, desertification, wetland destruction, severe erosion, and contamination (FAO, 2011, pp. 1-59).

By 2050, humanity could consume an estimated 140 billion tons of minerals, ores, fossil fuels, and biomass per year – three times its current appetite (UNEP, 2011). In particular, significant increases are expected regarding the global use of some critical commodities (National Intelligence Council, 2013). Specific examples of additional resource demands projected by various agencies include:

- **Food:** The UN Food and Agriculture Organization projects that, by 2050, the global demand for food will increase by 70% (Alexandratos and Bruinsma, 2012, pp. 59-64). Additionally, higher production of biofuels will increase competition for land.

- **Water:** Global water withdrawals have tripled in the last 50 years while the reliable supply of water has stayed relatively constant during the same period. Global growth in water demand for all uses is projected to increase about 50% by 2050, leaving three-quarters of the global population facing freshwater shortages. Countries already facing serious water stress include those of the Middle East, South Asia, and East Asia (FAO, 2011).
- **Energy demand:** According to the International Energy Outlook published by the U.S. Energy Information Administration, by 2040, global energy demand is projected to grow by 30% based on a 2012 baseline. Electricity demand is projected to increase by 69% with half of the additional demand supplied by renewable sources. Natural gas consumption will increase about 48% while the use of petroleum liquid fuels will increase by about 34%. Coal consumption is projected to increase by about 17%. The increases will mainly occur in the developing world, driven by long-term growth in economies and populations. More than half of the total world increase in energy consumption will occur in developing Asia (EIA, 2016a, pp. 7-12).
- **Energy access:** Despite overall growth, energy consumption per capita globally remains grossly unequal, although developing countries are catching up. Some 2.7 billion people around the world still rely on traditional biomass fuels for cooking and heating (resulting in about 3.5 million deaths annually from indoor air pollution). About 1.3 billion people do not have any access to electricity (International Energy Agency, 2016b). Parts of the developing world (China, India, and Indonesia, in particular) are undergoing rapid industrialization, urbanization, population growth,

and rising consumer demand. China and India's burgeoning manufacturing sectors have produced one of the biggest historical increases in power generation capacity but this has come at a huge cost. Air pollution currently claims about 5.5 million lives a year globally, making it the world's fourth-leading cause of death. India and China account for about 55% (3 million) of this total (Procopiou, 2016).

- **Minerals:** The amount of minerals, ores, fossil fuels, and biomass consumed globally per year could triple by 2050 (UNEP, 2011). Zinc, lead, and particularly copper and nickel mining will be affected by declining ore grades, as will precious metals such as gold and platinum. Falling ore grades may lead to an exponential increase in the amount of energy and water needed to extract metals. Recycling is likely to play an increasingly important role, but the development of the recycling industry over the past decades has been disappointing when compared with the conventional extraction of new ore.
- **Rare earth elements (REEs):** REEs include a group of 17 chemical elements that, because of their geochemical properties, are typically dispersed and difficult to find in concentrations that are feasible to mine. REEs are becoming increasingly integrated in new technologies, especially within the clean energy, military, and consumer electronics sectors. As each of these sectors continues to grow, so does the industry's demand for REEs. The current state of global REE supply and demand is reasonably well balanced, but that balance will likely change in the near future. For instance, the supply of dysprosium, an element used in the production of magnets for wind turbines and electric vehicles, is projected to meet only 15–18% of demand by 2025

and only 10% by 2040 (Moss *et al.*, 2011, pp. 80-83). Additionally, there is a lack of security in the REE supply chain. In 2011, over 95% of global REE supply originated from China, the largest user of REEs at 65% of total demand. The U.S.A. is the next largest user, at 15% of total demand. In 2010, China announced a 40% reduction in exports of REEs and future reductions are expected (Hatch, 2012).

- **Climate change:** The persistent accumulation of greenhouse gases in the atmosphere has been ongoing for decades and greenhouse gas levels are approaching thresholds that climatologists believe can disrupt the Earth's climate. They project that the warming climate will affect weather patterns, sea levels, community resilience, ecosystem health, and agricultural productivity. While some individuals remain unconvinced by the source and magnitude of the climate change issue, it must now be regarded as a social and business fact that is affecting billions of decisions around the world. The U.S.A. has historically lagged behind other nations with respect to the importance of climate risks and the need to take action, but that trend appears to be changing rapidly. A 2015 survey by the University of Michigan's Center for Local, State, and Urban Policy found that more than 70% of Americans now believe that climate change is real, and more than half of the world views the issue as the most serious global threat to humanity (Borick *et al.*, 2015). Another noteworthy aspect regarding the climate change issue is the perspective of millennials who tend to view climate change with the same degree of concern that baby boomers viewed the prospect of nuclear war.

## Growing uncertainty from increasing regulations and economic challenges

> Our plans miscarry because they have no aim. If one does not know to which port he is sailing, no wind is favorable.
>
> Lucius Annaeus Seneca

Uncertainty, associated with future responses to the growing constraints, is creating an extraordinary amount of constraint headwinds. The issues tend to be extremely complex and can impact diverse stakeholders through a wide range of economic, social, and environmental aspects. It is difficult to reach broad consensus on responses because they are dependent on the values and priorities of the individuals and organizations advocating for change. For instance, with respect to climate change, a complex web of policy measures focused on reducing fossil fuel use have already been implemented and many more have been proposed. The approaches differ greatly and include a wide range of measures, including carbon taxes at various levels, cap and trade systems, and carbon emission limits.

Effective strategies for responding to these measures vary greatly depending on the option implemented and the various levels of tax, emission limits, and so on. Additionally, many measures associated with land use are being actively promoted, including stopping deforestation, improving soil management, restoring degraded lands, and improving waste management practices. The extreme levels of complexity associated with these issues, and the various options for addressing them, has made consensus building difficult and produced considerable uncertainty regarding future constraints. Nothing stifles business activity and investment more than uncertainty. Consequently, some organizations have advocated for global standards that can be broadly implemented to reduce complexity and uncertainty and facilitate future planning. Even some of the world's largest oil and gas companies have publicly announced their support for a global carbon tax system, even though such a tax would impact them financially (Bloomberg, 2015).

Today's society is consuming and degrading resources of all types at rates that are not sustainable, and future generations will be saddled with overwhelming problems if a course correction is not implemented soon. In addition to the formidable constraints facing global social and environmental systems, the global economic system faces considerable uncertainty. The expanding global debt bubble is not sustainable in the long term and future generations will be saddled with this debt load and have few reasonable prospects for ever paying it back. Countries all over the world continue to leverage debt in hopes of stimulating their economies. Global debt of all types grew by US$57 trillion from 2007 to 2014 to a total of US$199 trillion (Dobbs *et al.*, 2015). That represents a ratio 300% higher than the world's GDP. The rising debt load, Dobbs *et al.* noted, "poses new risks to financial stability and may undermine global economic growth."

With declining availability of resources and increasing demand from population growth and an expanding middle class, higher prices for food, energy, metals, lumber, and water are inevitable. Resource-consuming countries face increasing trade imbalances with economies that are rich in resources. This could raise the specter of political tension between the two. Some resource-intensive industries, such as steelmaking, foundries, and heavy manufacturing, relocated most operations to developing countries with less stringent regulatory requirements. The relatively high-paid middle-class jobs associated with these industries often went with them. For instance, in the spring of 2016, the India-based steelmaking conglomerate, Tata, announced that it planned to sell its steelmaking operations in the U.K. (although it didn't actually have any potential buyers lined up) and keep the bulk of its steel production capabilities operating in China and India. This move would potentially eliminate thousands of jobs for U.K. steelworkers and the businesses that support the sector (BBC News, 2016).

Although the factors that generate constraint headwinds have cost global economies trillions of dollars in lost opportunities, businesses have historically struggled to take effective actions to address them. The levels of uncertainty associated with future headwind constraints are overwhelming. Markets usually fail to account for the monetary value of indirect social and environmental costs such as pollution and public health and safety. Since the monetary benefits are not widely accepted, businesses often have difficulty justifying significant investments in protecting and improving these aspects of their enterprises.

## Emphasis on short-term financial performance

Investors have placed considerable pressure on businesses to achieve short-term financial performance rather than creating long-term, sustained value for shareholders and stakeholders. Capital markets have evolved to emphasize trading over investment and placed substantial emphasis on quarterly earnings as the principal measure of corporate performance. Businesses have increased their commitments to social responsibility in recent years but pressure to achieve short-term financial results has tended to limit their direct involvement in the wellbeing of communities and the environment. These factors do not typically have a direct impact on the short-term financial parameters that business leaders are judged on.

Consequently, businesses tend to focus largely on managing factors that directly impact financial performance such as productivity, cost control, and revenue generation.

This problem has become so commonplace that, in the spring of 2016, Larry Fink, the chief executive at BlackRock, the world's biggest investor with US$4.6 trillion, sent a letter to chief executives at S&P 500 companies and large European corporations. Here are a few of his comments from that letter (Turner, 2016):

> Many companies continue to engage in practices that may undermine their ability to invest for the future. Dividends paid out by S&P 500 companies in 2015 amounted to the highest proportion of their earnings since 2009. As of the end of the third quarter of 2015, buybacks were up 27% over 12 months. We certainly support returning excess cash to shareholders, but not at the expense of value-creating investment. We continue to urge companies to adopt balanced capital plans, appropriate for their respective industries, that support strategies for long-term growth ...
>
> Annual shareholder letters and other communications to shareholders are too often backwards-looking and don't do enough to articulate management's vision and plans for the future. This perspective on the future, however, is what investors and all stakeholders truly need, including, for example, how the company is navigating the competitive landscape, how it is innovating, how it is adapting to technological disruption or geopolitical events, where it is investing and how it is developing its talent.

## The need for regulatory reform

Due to the inability of markets to effectively address the many system failures and consequences associated with industrialization, governments continue to develop regulatory measures designed to protect people and the environment. Most ethical companies appreciate a reasonable amount of thoughtful regulation and enforcement because it tends to thin out unethical competitors and "level the playing field." However, regulations also have a tendency to accumulate over time and agencies rarely go back and delete or update requirements that are no longer appropriate.

Unfortunately, government regulation has contributed greatly to increasing constraint headwinds by implementing an exhaustive system of requirements, reports, and incentives that continuously grow in volume and complexity. Regulators have good intentions when proposing new rules, such as increasing worker safety or protecting the environment. However, policy-makers typically view each regulation on its own, paying little attention to the rapid build-up of rules, many of them outdated and ineffective, and how that regulatory accumulation hurts economic growth.

**Case study: Regulations impede dry cell battery recycling**

Many governments have begun mandating battery recycling in recent years. This seems like a good idea given that the U.S.A. alone discards some 3 billion dry cell batteries each year, and given that, when they degrade in landfills, batteries release a variety of potentially harmful chemicals such as heavy metals. Unfortunately, industry response to the mandates has been slow because of some dysfunctional regulatory dynamics. Due to their complex chemistry, regulators have classified used batteries as hazardous waste to ensure they are disposed of correctly. Therefore, from a regulatory perspective, a company that recycles used batteries would not only be in the battery production business, but they would also be in the hazardous waste management business.

The rules for managing and disposing of hazardous wastes tend to be extremely technical and complex, and they pose a huge regulatory burden for the companies that have to deal with them. In comparison, companies that produce new batteries from scratch are not considered to be in the hazardous waste management business. They use the same potentially harmful materials to produce their products, but since their raw materials do not come from used batteries, they are not subject to hazardous waste regulations. New battery producers obtain their virgin chemical inputs from mining operations and chemical processing plants instead of using chemicals from resources that are already in the value chain. This dysfunctional regulatory system is largely responsible for the slow rate of battery recycling and the consequent overconsumption of raw materials and excessive waste generation.

Economists have estimated that market distortions resulting from a heavily regulated economy result in slower economic growth (2–3% slower than a moderately regulated economy). Between 1949 and 2005, the accumulation of federal regulations slowed U.S. economic growth by an average of 2% per year. A 2005 World Bank study found that a 10% increase in a country's regulatory burdens slows the annual growth rate of GDP per capita by 0.5%, resulting in thousands of dollars in lost GDP per capita growth in less than a decade (McLaughlin and Greene, 2014).

Some regulatory initiatives have been implemented using practices that are far less disruptive to economic development than the examples provided above. The Montreal Protocol provides an excellent example of innovative public–private sector collaboration that addressed an environmental issue with effective regulatory measures that had minimal impacts to the economy.

## Case study: The Montreal Protocol innovation in regulation

The Montreal Protocol is one of the most successful and effective environmental treaties ever negotiated and implemented. It was designed to reduce the production and consumption of ozone-depleting substances in order to reduce their abundance in the atmosphere, and thereby protect the Earth's fragile ozone Layer. The original Montreal Protocol was agreed on September 16, 1987 and implemented on January 1, 1989 (The Conversation, 2012).

The Montreal Protocol broke new ground in both its negotiation and in its construction. It was ratified, or accepted, by all 197 UN member states, a world first for any treaty. Such acceptance reflected the strong global commitment to this treaty.

Several factors contributed to its success, including:

- **Cooperation:** Unprecedented level of cooperation and commitment shown by the international community
- **Leadership and innovation:** Negotiations focused primarily on leadership and innovative approaches
- **Flexibility:** The protocol was designed to be a highly flexible instrument which could be adjusted to increase or decrease controls as the science became clearer
- **Nonpunitive compliance:** The Protocol provided a stable framework that allowed industry to plan long-term research and innovation
- **Industry participation:** Chemical companies developed innovative new chemistries that led to reasonably priced formulations with no ozone-depleting potential for use in the refrigeration and air-conditioning sectors

Since the Protocol came into effect, the atmospheric concentrations of the most important chlorofluorocarbons and related chlorinated hydrocarbons have either leveled off or decreased. A 2015 report by the EPA estimates that the protection of the ozone layer under the treaty will prevent over 280 million cases of skin cancer, 1.5 million skin cancer deaths, and 45 million cataracts in the U.S.A. (EPA, 2017a).

The public's understanding and concerns regarding constraint headwinds are growing, and so are their demands for products and services that are less resource intensive and less impactful on people, communities, and the environment. As a result, public concerns over business externalities will affect companies' "license to operate" much more frequently in the future. In some sectors, constraint headwinds have intensified to levels that exceed the technical, organizational, and/or economic capacities required to develop effective solutions. Innovators are challenged to find cost-effective solutions because of formidable technical and/or economic hurdles. When these conditions exist, the constrained organizations become vulnerable to competing alternatives. For many years, the coal sector has repeatedly dealt with a wide variety of constraint headwinds associated with the negative impacts that coal mining and use poses to community and environmental resources. These headwinds have dramatically impacted the industry because addressing the constraints has required major modifications to mining methods and expensive measures for controlling air emissions. The following case study describes the evolution of the coal sector's constraint headwinds and the measures used to address them in more detail.

**Case study: Constraint headwinds challenge the coal sector**

Throughout coal's history, those engaged in the production and utilization of this enabling energy source have repeatedly been confronted by a growing and diverse mix of constraint headwinds. Constraints have surfaced regularly over the span of many decades to challenge the viability of the sector. The coal sector, and the agencies that regulate it, have responded to each constraint with a combination of policies and technical solutions to address the issues at hand. To date, the solutions required for compliance have been viable from both technical and economic perspectives. However, new challenges associated with $CO_2$ emissions and their effects on the Earth's climate are proving to be particularly problematic for this sector.

During early industrialization, coal filled a critical need as fuel to produce heat and electricity that contributed greatly to industrial productivity and community development. However, along with the benefits realized from coal mining and utilization came some undesirable side-effects, including:

- Catastrophic accidents from mine failures
- Land degradation and water pollution from mining operations
- Air pollution resulting from fuel combustion

In the early stages of coal development, the side-effects and impacts were not widespread. They occurred in relatively isolated geographies and were sources of concern but considered acceptable by most stakeholders. Accidents were perceived to be tragic but controllable, air emissions quickly dissipated and land disturbances were small and relatively isolated. So, the ends (greater industrial productivity, economic growth, and community access to energy) appeared to justify the means.

As coal production and utilization expanded to meet the growing demand for energy, so did the magnitude and complexity of coal's side-effects and impacts. Most of the impacts were readily visible and society found them to be objectionable. In response to the growing problems associated with coal mining and utilization, a plethora of policies and innovations emerged to improve safety, restore disturbed lands, and control emissions. The frequency and impacts of mining accidents, especially from underground mines, led to aggressive policy measures in the developed world. Measures to improve safety and prevent incidents such as mine collapses and explosions were progressively implemented. The movement to improve mine safety peaked in the U.S.A. in 1977 when the U.S. Department of Labor created the Mine Safety and Health Administration (MSHA) with the specific purpose of reducing deaths, injuries, and illnesses in the nation's mines (MSHA, 2017a).

MSHA is required to perform annual inspections at all mines (four times per year for underground mines and twice per year for surface mines). MSHA also strengthened rights for miners and protected them from retaliation when they exercised their rights. Mandatory miner training programs were established along with the establishment of mine rescue teams (MSHA, 2017b).

The rapidly expanding acreage disturbed by mining operations was also a major source of public concern. Abandoned mine lands are considerably less productive than undisturbed lands. Large piles of steeply sloping earth were left, without taking steps to preserve and replace topsoil or revegetate the landscape. Natural drainage patterns were disrupted and the steep slopes and poor vegetative cover produced increases in run-off and erosion, resulting in severe pollution to nearby streams and rivers. In some cases, acid-producing geologic materials were left on the surface that generated acid mine drainage. These acidic materials were often toxic to vegetation and left large acreages barren while also polluting water bodies.

In 1977, the U.S. Congress passed the Surface Mining Control and Reclamation Act (SMCRA) to regulate surface mining activities and ensure the reclamation of coal-mined lands. The Office of Surface Mining Reclamation and Enforcement was established in the Department of the Interior to provide guidance, policies, and procedures required for all coal surface mining and the surface effects of underground mining. As a result of SMCRA implementation, mine operators are required to minimize disturbances and adverse impact on fish and wildlife, and to restore impacted land and water resources (OSMRE, 2017).

Air emissions associated with coal combustion have been a concern from the beginning. Initial concerns focused on the visible particulate matter that left unsightly soot accumulation all over the communities that burned coal. Advances in air chemistry revealed a variety of other problematic air pollutants, including sulfur dioxide (the primary contributor to acid rain), nitrogen oxides, mercury, and dozens of other substances that impacted large regions. Pollution control standards have been progressively developed to address this growing list of pollutants, and power utilities have responded with various pollution control technologies that remove the harmful constituents.

Addressing the expanding list of constraint headwinds has added considerable expense to the business of producing and utilizing coal. Most measures that have been implemented to mitigate coal's impacts have been costly controls and treatments that address problems after the fact. More cost-effective measures based on innovations in prevention and efficiency have not surfaced to date. In spite of the costly control measures, the coal sector has always managed to absorb the additional costs and remain economically competitive – until recently.

The newest and most challenging consequence of coal combustion has emerged from advances in climate science. Scientists now believe that $CO_2$ emissions from coal combustion (along with other fossil fuels) are a major source of greenhouse gases that are contributing to climate change. A push for new regulations and taxes has surfaced in recent years that is driving the coal sector to manage $CO_2$ emissions as a pollutant in many parts of the world. No cost-effective measures currently exist that can reduce the $CO_2$ emissions to acceptable levels and markets for recovered $CO_2$ are limited. Consequently, countless companies that have historically depended on the production, shipping, and utilization of coal have gone out of business and many others are floundering. The additional cost of controlling $CO_2$ emissions has left coal vulnerable to replacement with alternative power sources. Interest in wind, solar, geothermal, biomass, and nuclear energy has increased. However, natural gas popularity has grown even more dramatically as a result of well stimulation techniques that have increased production and lowered costs. Natural gas emits about 40% less $CO_2$ emissions per unit of energy than bituminous coal (EIA, 2017).

As shown in Figure 3.1, consumption of steam coal used for electricity generation in the U.S. electric power sector fell 29% from its peak in 2007 to 2015. Consumption fell in nearly every state, rising only in Nebraska and Alaska over that period.

Based on the long history of constraint headwinds associated with the coal sector, one might assume that its future as an energy source is in jeopardy. The coal sector needs to overcome a major headwind associated with $CO_2$ emissions and many experts believe that the $CO_2$ obstacle will deal a death blow to the coal sector.

However, proclamations of the coal sector's death may be a bit premature given that proven technologies for capturing $CO_2$ are now available. These technologies are not currently cost-effective

**FIGURE 3.1 U.S. power sector coal demand, 2007–15 (million short tons)**

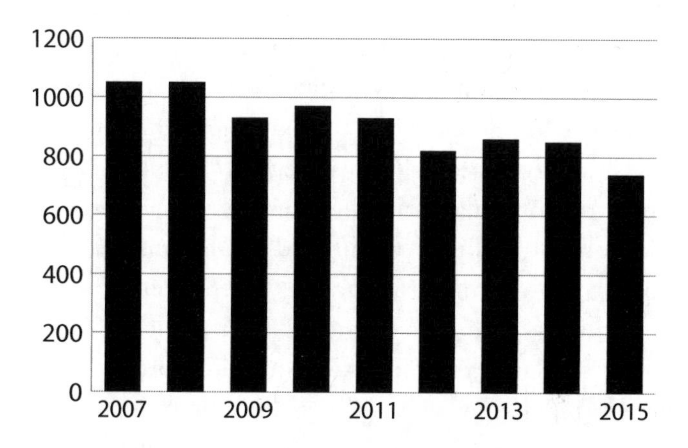


Source: EIA, 2016b.

but, if innovations emerge that can lead to economical uses for the captured $CO_2$, coal may still be an important energy source for decades to come. The notion that such innovations may be commercially deployed is far from outlandish. Nature, through photosynthesis, has been using $CO_2$ for millions of years as a feedstock to make immeasurable quantities of food, fiber, fuel, and lumber. Mankind may be able to develop technology that will make valuable products from the $CO_2$ as well. Hundreds of institutions are currently researching methods for producing valuable products from $CO_2$ and considerable progress is being made. This includes a US$20 million XPrize competition currently underway to develop breakthrough technologies that will convert $CO_2$ emissions into valuable products (XPrize, 2017). Coal's future will depend in large part on the ability of innovators such as these to develop effective innovations and markets for the products.

## Market failure intensifies constraints

Markets have usually failed to account for indirect social and environmental costs such as pollution and public health and safety. As constraints grow and intensify, various stakeholders have reacted with proposals to address them through a combination of policy and/or technology-based solutions. Regulators and social interests have attempted to protect community and environmental resources with additional subsidies, penalties and regulations.

As constraints continue to magnify, governments respond with more regulations to address them and the complexity and cost of compliance escalates as well. Establishing the proper amount of government regulation that still enables innovation, economic development, and wealth creation across all segments of society can be challenging. The net effect tends to be stronger and stronger headwinds for organizations that deal reactively with the growing constraints. To date, many of these measures have been reactions to issues after they have become crises. They tend to focus on the outcomes and symptoms of complex problems as opposed to addressing the root causes. More effective measures have been implemented by some forward-thinking organizations that address the root causes of constraint headwinds, as discussed in the following chapter.

4

# Proactive approaches emerge

> When everything seems to be going against you, remember that the airplane takes off against the wind, not with it.
>
> Henry Ford

Organizations frequently assume that constraint headwinds originate mostly from external factors beyond their influence. This assumption usually occurs when organizations delay taking action until after the growing constraints have escalated to a point where market conditions have declined or governments have intervened with regulatory requirements. The affected organizations often believe that their only recourse for responding to the regulations is through reactive compliance measures. However, some industry leaders have recognized that the growing headwinds they face are not solely the result of uncontrollable market conditions and expanding regulations. They acknowledge that their organizations contribute to constraint headwinds through various aspects of their practices, processes, products, and systems. Some proactive leaders have been able to successfully minimize their constraint headwinds by addressing problematic aspects of their operations at a root cause level. Doing so has frequently improved their effectiveness and provided a source of competitive advantage.

## Addressing the root causes of constraint headwinds

The vast majority of organizations that experience constraint headwinds contribute a great deal to their intensity through the ways they manage their processes, products, and interactions with suppliers, customers, communities, and the environment. Four basic categories of root causes are common in the systems of organizations that face considerable constraint headwinds:

1. Wasteful practices
2. Deficient processes and products
3. Dysfunctional interactions
4. Negligent decisions

The characteristics of these root causes, and the impacts they have on constraints, can be summarized as follows:

### 1. Wasteful practices

- **Characteristics:** Practices that result in inefficiency, hazards, accidents, attrition, defects, by-products, and excessive resource consumption

- **Impacts:** Injuries, diseases, damages, emissions, and wastes; and consequential wasted talent, depletion of resources, degradation of resources, pollution, and escalating costs

2. *DEFICIENT PROCESSES AND PRODUCTS*

- **Characteristics:** Processes and products with defects that negatively affect quality, safety, productivity, efficiency, and effectiveness of people, communities, businesses, resources, and the environment
- **Impacts:** *Wastefulness* that leads to reduced quality, productivity and availability of community, environmental, and businesses assets; and consequential reduced quality of life for people

3. *DYSFUNCTIONAL INTERACTIONS*

- **Characteristics:** System aspects are not configured and managed in ways that maximize the benefits of exchanges between suppliers, designers, producers, customers, communities, and the environment.
- **Impacts:** *Deficiency* and *wastefulness* regarding processes, products, methods, people, machines, materials, and energy

4. *NEGLIGENT DECISIONS*

- **Characteristics:** Choices that are unethical, irresponsible, and/or not compliant with regulations and policies; causing harm to people, organizations, communities, and the environment

- **Impacts:** Compliance failure; lawsuits, workforce attrition, declines in reputation and license to operate; damage to community, environmental, and personal resources; *dysfunction*, *deficiency*, and *wastefulness*

Not only do these dynamics contribute to constraint headwinds, but they also limit an organization's effectiveness and competitiveness. Given that most organizations contribute to the constraint headwinds they face through their unsustainable practices, processes, products, and systems, it stands to reason that these same organizations can take proactive measures to mitigate the constraints by improving these aspects.

## Dealing with deficiency and dysfunction

Deficiencies associated with production processes and the performance of products can create a plethora of problems that are not good for businesses and their relationships with communities and the environment. Defects and hazards can occur throughout an organization's value chain that can lead to accidents, injuries, inefficiency, wastes, costs, emissions, pollution, and resource depletion. They can also contribute to strained relationships with employees, community leaders, and other stakeholders. At the end of a product or component's life-cycle, they often become waste and require disposal, thus ensuring that the valuable resources that went into product fabrication are removed from the enterprise value chain forever.

Industrial systems have historically been linear, sloppy, and dysfunctional, as the various enterprise components tend to be designed and developed in isolation with little regard for how they will interact with other components. Measures taken to improve the performance of one system component regularly create problems for other components. For example, increasing the production schedule of an operation by adding another shift can result in maintenance problems if the maintenance schedule is not increased proportionately. These types of dysfunctional interactions tend to become common as those responsible for individual components strive to improve performance without regard for their interactions with, and impacts on, other components. Dysfunctional interactions can occur throughout the value chain, including various aspects of design, supply chain, logistics, production, maintenance, distribution, and customer support. Changes made to any of these aspects can, and frequently do, affect the performance of other aspects.

By implementing measures that address the sources of deficiency and dysfunction, organizations can greatly reduce their exposure to constraint headwinds. This includes all types of measures to improve the life-cycle of processes and products by preventing waste, defects, hazards, and inefficiencies. It also includes measures that optimize the interactions between system components in ways that are mutually beneficial.

## Prevention-based strategies

With better understanding of the factors that comprise the root causes of constraint headwinds, forward-thinking organizations are developing strategies that address the problems and constraints more proactively. Instead of seeking to simply comply with regulatory requirements, they implement prevention-based approaches focused on improving processes and products through measures that eliminate problems at their root causes. Prevention-based strategies are increasingly being recognized as more effective approaches for addressing a wide range of problems when compared with approaches that attempt to deal with problems after they already exist. Many organizations have successfully implemented prevention strategies in ways that improve their efficiency, their effectiveness, and their bottom line These strategies regularly lead to substantial improvements in the safety, quality, productivity, and efficiency performance of processes and products by preventing a variety of accidents, illnesses, wastes, costs, emissions, and other problems. The quest for improvements in these dimensions has become a principal driver of innovation in the industrial world and has led to many technology breakthroughs.

## Accident prevention

Government-imposed safety standards and regulations focus primarily on the safety of work conditions, whereas approximately 80–95% of all accidents are triggered by unsafe behaviors. Some proactive organizations have achieved considerable improvements in their safety performance by implementing accident prevention strategies that include a combination of training and supervision to ensure that personnel consistently perform their duties safely. For example, DuPont developed its "Safety Training Observation Program" (STOP) to help organizations prevent accidents and injuries by increasing safety awareness associated with preventing unsafe behaviors. Previous efforts to improve safety performance had focused predominantly on addressing unsafe conditions. These behavior-based safety programs drive workers to share in the responsibility of ensuring workplace safety (Byrd, 2007, pp. 58-64). A case study of the DuPont STOP program is provided in the section "Addressing the root causes of wastefulness" in Chapter 10.

Officials at the National Institute of Safety and Health (NIOSH) recognized the superior safety benefits that could be achieved with prevention strategies and initiated their "Prevention Through Design" (PtD) initiative in 2007 (National Institute of Safety and Health, 2017). The mission of the PtD initiative is to prevent or reduce occupational injuries, illnesses, and fatalities through the inclusion of prevention considerations in all designs that impact workers. PtD practitioners evaluate the root causes of potential risks associated with processes, structures, equipment, and tools. They take steps to "design out" or minimize hazards and risks associated with a variety of processes, including construction, maintenance, decommissioning, and disposal or recycling of waste material. Key elements of PtD include:

- Eliminating hazards and controlling risks to workers "at the source" or as early as possible in the life-cycle of workplaces
- Including design, redesign, and retrofit of new and existing work premises, structures, tools, facilities, equipment, machinery, products, substances, work processes, and the organization of work
- Enhancing the work environment through the inclusion of prevention methods in all designs that impact workers and others on the premises

NIOSH now provides ten engineering textbooks and 25 consensus standards based on PtD standards.

## Prevention of waste and pollution

Although new challenges continuously surface as industrialization expands, entrepreneurs have often been able to address growing constraints with innovative approaches that conserve resources and make productive use of by-products and wastes. The agricultural sector has implemented a variety of conservation practices to better manage run-off, prevent erosion, and ensure the long-term productivity of land resources. Ford is famous for shipping the Model A truck in crates that it dismantled at the factory destination, then used the crate lumber to make the vehicle's floorboards. Rudolf Diesel was able to take a distillate by-product left over from producing gasoline and use it for fuel (commonly referred to as diesel fuel today) in his (at the time) revolutionary new compression-ignition engine.

In the mid-1970s through to the 1990s, considerable innovation took place regarding the concept of "pollution prevention." The basic concept behind pollution prevention is that wastes and emissions are an outcome of deficient products or processes where valuable raw materials are converted to forms that have limited or even negative value. Controlling, treating, and recycling the wastes and emissions frequently adds considerable expense and complexity to processes and products. Instead of focusing on how best to treat or recycle emissions and wastes after they are generated, pollution prevention advocates direct their attention to the root cause. They develop and implement processes, machines, materials, and chemicals that are inherently safer, more productive, and more efficient. They optimize systems to ensure that raw material inputs culminate in value-added products instead of costly emissions and wastes. In 1975, 3M initiated its "Pollution Prevention Pays" program as a more cost-effective response to the environmental legislation and regulations of the early 1970s. A case study of this program is provided in the section "Addressing the root causes of wastefulness" in Chapter 10.

Pollution prevention measures usually require less raw material inputs and generate less emissions and wastes. These improvements usually result in considerable cost savings, particularly when compared with conventional treatment and control approaches that add to overhead costs. For instance, less hazardous coating technologies are now available that can be applied as a dry powder, or as an aqueous solution where water is substituted for hazardous solvents in the formulation. By removing the hazardous solvent from the paint formulation, users also eliminate a major source of air pollution and hazards – the evaporating solvent.

## Prevention of social problems

Prevention strategies are increasingly being used to address a diverse mix of social problems. Strategies that keep crime from being committed, a substance user away from dependence, a child safe from harm, or a student in school are considerably more effective than attempting to rebuild lives scarred by jail, addiction, maltreatment, or a failed education. When one social problem (e.g., child abuse) is prevented, several other social problems are usually prevented as well. Attention to prevention in social and healthcare services is unmistakably on the rise. For example, the U.S. 2010 healthcare reform bill, the Patient Protection and Affordable Care Act, created the national Prevention, Health Promotion and Public Health Council, provided grants to small employers to create wellness programs, mandated the disclosure of nutrition content in chain restaurants, and required insurers to cover key evidence-based clinical preventive services. Countless other prevention strategies have been implemented through programs focused on preventative medicine, vaccines, crime prevention, drug abuse prevention, and so on.

# 5

# Transitioning from linear flow to a circular economy

> When the winds of change blow, some people build walls and others build windmills.
>
> Chinese proverb

Early industrialization was based primarily on a linear approach to resource use. This approach is still in practice in many sectors and includes three basic steps:

1. Extraction → 2. Utilization → 3. Disposal

Raw materials and energy were fed into processes and transformed into products that were shipped to customers. By-products of production such as emissions and wastes also exited the processes and accumulated in land, water, and air assets, resulting in considerable degradation of these resources. At the end of their useful life, most products became waste as well and were discarded, permanently removing important resources from the value chain. The declining quality and availability of resources negatively impacted the communities that depended on them.

As resource availability has declined and the cost of raw material extraction has increased, alternative approaches based on a circular flow of materials and energy have emerged. A great deal of potential value can be gained by optimizing product design and production processes to achieve multiple cycles of disassembly, refurbishment, and reuse. Circular flow is achieved by connecting an organization's value chain end to end, from start to finish.

For many years, consumption of raw materials could be directly correlated with economic growth. However, that relationship has begun to decouple in recent years as a circular economy has emerged. Entire industries have surfaced that include a combination of materials, processes, and systems that enable reusing, recycling, and remanufacturing of worn materials and components. The global economy today generates 50% more economic value from a ton of raw materials than it did in the 1980s (OECD, 2015, pp. 70-89).

## Materials recovery and recycling

The increasing demand for affordable materials spurred by industrialization led entrepreneurs to develop businesses focused on recovering materials from various sources of scrap, by-products, and wastes. Early efforts focused on ferrous scrap metals because they were cheaper to acquire than virgin ore. Railroads both purchased and sold scrap metal in the 19th century, and the growing steel and automobile industries purchased scrap in the early 20th century. Many secondary goods were collected, processed, and sold by peddlers who scoured dumps and city streets for discarded machinery, pots, pans, and other sources of metal. By World War I, thousands of such peddlers roamed the streets of American cities, taking advantage of market forces to recycle postconsumer materials back into industrial production (Zimring, 2005).

Recycling was a highlight throughout World War II as financial constraints and significant material shortages made it necessary for countries to reuse goods and recycle materials. Massive government promotional campaigns were carried out in every country involved in the war. They urged citizens to donate metals and conserve fiber, as a matter of patriotism. After the Japanese attack on Pearl Harbor, the U.S. government organized a major conservation and recycling effort. Americans were asked to salvage and collect a long list of materials, including paper, aluminum, tin, iron and steel, rubber, silk, and cooking fat. Cities and states were given quotas and families were involved with conservation and recycling of goods. Even children were included in the recycling effort as they participated in drives to collect scrap metal, used tires, and paper for the war effort. Contests were held to meet established quotas and some companies sponsored drives and offered prizes. The scrap drives received a great deal of publicity, especially early in the war, as newspapers reported on the quantities of material collected. It was something that the average citizen could do to support the country at a time of great national need (Rockoff, 2000, pp. 25-26).

Following World War II, interest in material salvage and recycling declined. This was partly due to the reduced sense of need and urgency associated with the war ending. Additionally, the emergence of a growing multitude of convenient throwaway items also contributed. The disposable items were cheap, often made out of plastics produced from widely available petroleum-based materials. In the 1970s, oil prices increased dramatically and incentivized considerable investment in recycling. Scrap recycling is currently a

US$90 billion industry in the U.S.A. that annually transforms more than 135 million metric tons of obsolete materials from consumers, businesses, and manufacturers into useful raw materials that are essential to the manufacturing of new products (Institute of Scrap Recycling Industries, 2017). The rates at which materials are currently recovered varies greatly from country to country and between the developed and developing worlds. In spite of many advances in recycling practices and technology, recycling has plateaued in the U.S.A. in recent years. Low landfill fees and a fragmented waste management system have kept the U.S. material recovery rate (recycling and composting) at around 34% for two decades. This wastefulness comes at considerable cost as the U.S. wastes US$11.4 billion worth of recyclable containers and packaging annually (OECD. Stat, 2015).

The U.S. materials recovery rate of 34% is far lower than most developed countries, such as Germany (64%), Korea (59%), Austria (58%), Belgium (55%), and Switzerland (54%). In contrast, many countries in the developing world still do not have effective materials recovery systems in place. For instance, China recovers less than 10% while Mexico recovers about 5% and Turkey only recovers about 1% (OECD.Stat, 2015).

In nearly 50 countries, including the entire European Union, a policy called "extended producer responsibility" shifts the burden of waste disposal from taxpayers to companies. Producers are required to design, manage, and finance programs for end-of-life management of their products and packaging as a condition of sale. Programs cover all products in a given category, including those from companies that are no longer in business and from companies that cannot be identified. Results associated with these programs are inconsistent in terms of costs and effectiveness, but they have proven to be effective for increasing recycling rates and saving local municipalities money (The Economist, 2015a).

The unfavorable economics associated with some recycled materials have become more challenging recently. Falling oil prices have lowered the price of virgin plastic to a point where using recycled commodities does not necessarily help a company's bottom line. Some companies have committed to using more recycled materials in their products but have struggled to acquire adequate supplies of feedstock. Coca-Cola committed to using at least 25% recycled plastic in its containers by 2015, but revised this downwards due to a limited supply and unfavorable economics. Walmart is struggling to find enough material to meet its goal to use 3 billion pounds of recycled plastic in its packaging and products by 2020. In order to boost the supply of recyclable feedstocks, some companies are starting to invest in recycling infrastructure. Walmart, Coca-Cola, and eight other big companies have created a US$100 million Closed Loop Fund, which offers zero- and low-interest loans to cities and recycling companies for everything from better bins to more efficient sorting plants. The loans are repaid with the profits from diverting waste from landfills and selling the materials (American Iron and Steel Institute, 2006).

In theory, most recycled materials should be cheaper than virgin commodities. However, this is frequently not the case because most products are not designed with recycling in mind. The concept of sustainable design aims to solve this problem, and was laid out in the book *Cradle to Cradle: Remaking the Way We Make Things* by architect William McDonough and chemist Michael Braungart (McDonough and Braungart, 2002). They suggest that every product (and all the packaging it requires) should have a complete "closed-loop" cycle mapped out for each component. They encourage the manufacture of products with the goal of either returning the materials to the natural ecosystem through biodegradation or by recycling the components and materials indefinitely.

Considerable resources are conserved through recycling. For example, every ton of steel that is recycled saves about 120 pounds of limestone, 1,000 pounds of coal and 2,500 pounds of iron ore (American Iron and Steel Institute, 2006). Not only are the materials preserved, the amount of energy required for recycling is considerably less than the energy required to produce materials from virgin inputs. The Institute of Scrap Recycling Industries provides the following estimates for energy savings associated with commonly recycled commodities (Institute of Scrap Recycling Industries, 2015):

- Aluminum recycling requires 92% less energy
- Copper recycling requires 90% less energy
- Plastic recycling requires 87% less energy
- Iron and steel recycling requires 56% less energy
- Paper recycling requires 68% less energy

Fossil fuels are the predominant source of energy used in producing these materials. Therefore, in addition to reduced energy consumption, proportional reductions in air emissions from fuel combustion are also achieved.

Recycled materials supply 40% of the global raw material needs. This US$200 billion market employs about 1.6 million people worldwide. Approximately 500 million tons (38%) of the total steel produced globally in 2008 were made from scrap metal (Bureau of International Recycling, 2017). Compared with other materials, aluminum is relatively simple and cost-effective to recycle. The U.S. aluminum industry pays out more than US$800 million per year for recycled cans. As a result, nearly 75% of all aluminum ever produced in the U.S.A. is still in use. The global recycling rate for aluminum is about 69% (The Aluminum Association, 2017).

## Construction and demolition debris

The USEPA estimates that about 170 million tons of construction and demolition debris, comprised of drywall, concrete, cardboard, metals, and plastic, is produced annually in the U.S.A. About 48% of this material is currently recovered and recycled compared with just 25% in 1996. These figures are strictly based on building construction waste, which doesn't take into account wastes from road or bridge construction (EPA, 2017b).

Asphalt pavement is America's most recycled and reused material. It is being recycled and reused at a rate of over 99% with over 62 million tons recycled in 2010. New asphalt now contains an average of 12% recycled content. The use of reclaimed asphalt pavement began in the 1970s and, by the 1980s, use of recycled asphalt had become common practice. In addition to recycling used asphalt, good progress has been made with incorporating by-products from other sectors into asphalt mixes (National Asphalt Pavement Association, 2010). Examples include:

- **Waste shingles:** Over 1 million tons of waste shingles are now blended into asphalt each year because the composition of the shingles actually improves asphalt quality.
- **Waste tires:** Adding ground tire rubber to asphalt at rates of up to 15% can contribute to improved rutting resistance, skid resistance, ride quality, and reduced pavement noise levels. It also increases pavement life by lessening brittleness and cracking (Walker, 2010).
- **Foundry sand:** Foundry sand is used to form molds for metal castings at about 3,000 active foundry operations in the U.S.A. In spite of recent innovations that increase sand recycling, waste sand is generated at rates of 6–10 million

tons per year. Foundry sand has regularly proven to be of equal or higher quality when compared with the virgin sand commonly used to make asphalt (EPA, 2007).

Approximately 200 million tons of waste concrete is generated annually in the U.S.A. from a combination of construction and demolition and public infrastructure projects. About 140 million tons of this material is now recycled as a growing number of transportation agencies are incorporating recycled concrete aggregate into new concrete mixes. Recycled aggregate has proven to be at least as durable as virgin aggregates and 15–20% cheaper. Additionally, the U.S. concrete industry currently incorporates a significant amount of industrial by-products into new concrete blends. Fly ash and blast furnace slag that would otherwise end up in landfills are now commonly blended with the cement used to make concrete. Fly ash is the waste by-product (15–20% by volume) left over from burning coal to produce electric power. Blast furnace slag is the waste by-product of iron manufacture (Construction and Demolition Recycling Association, 2014).

## Remanufacturing

Over the past several decades, a remanufacturing industry has emerged where used products are taken back, reusable components are salvaged and refurbished, and valuable materials are recovered. The refurbished components are then redistributed to customers who use them for another life-cycle while recovered materials are used to make new components and products. By utilizing materials and components that are already in the value chain, resource consumption is greatly reduced. Remanufacturing is a process that involves dismantling a product, restoring and replacing components, and testing the individual components and the whole product to ensure that it meets original design specifications. The performance after remanufacture is expected to be "like new" and the remanufactured product generally comes with a warranty. The refurbished components are resold to customers that benefit from another life-cycle of service with a much smaller investment of resources. The automotive sector estimates that components can be remanufactured with 86% less energy than new parts. Customers can purchase a wide range of remanufactured components (e.g., engines, transmissions, fuel pumps, water pumps, and alternators) for about 40% less cost than new (UNEP, 2013, pp. 187-189).

The U.S.A. and European Union account for the bulk of global remanufacturing activity and trade. Although Brazil, India, and China are developing their own remanufacturing industries in response to growing domestic demand, they also tend to restrict trade in remanufactured goods and related inputs (USITC, 2012). During the period from 2009-2011, U.S. production of remanufactured goods grew by 15% to over US$43 billion and supported 180,000 full-time jobs. The largest US remanufacturing sectors are aerospace, heavy-duty and off-road equipment, and motor vehicle parts. U.S. exports of remanufactured goods totaled US$11.7 billion in 2011, up 50% compared with 2009. Canada, the European Union, and Mexico are important markets for US exports of remanufactured goods (USITC, 2012, pp. 10-1 to 10-10).

## Bio-based materials

The recycling and remanufacturing systems described above can prevent waste by creating a circular flow of resources for multiple life-cycles. These systems are particularly effective when applied to nonrenewable resources such as metals, petroleum-based plastics, and building materials. Biological-based ("bio-based") materials provide another means of establishing circular flow by utilizing natural biological processes to produce materials and eradicate waste. By emulating nature (a process often referred to as "biomimicry"), bio-based materials are made from ingredients produced biologically in agricultural and forestry operations. They are usually biodegradable which allows them to be composted at the end of their useful life instead of taking up space in landfills (Sustainable Biomaterials Collaborative, 2017). Bio-based materials have achieved considerable market penetration in a variety of sectors, including the chemicals, packaging, cosmetics, and plastic/polymer industries. Examples of bio-based products include:

- Plastics
- Paints and coatings
- Fuel additives
- Inks
- Sorbents
- Landscaping materials
- Adhesives
- Lubricants
- Solvents and cleaners
- Fibers and packaging
- Construction materials
- Compost and fertilizer

The global market for bio-based materials is growing rapidly. It was estimated to be US$49 billion in 2015 and is projected to exceed US$84.3 billion by 2020. Most of the growth will probably occur in specialty chemicals such as adhesives, surfactants, solvents, and plastic polymers. Specialty chemicals alone constitute about 60% of the total value of all bio-based chemical production (Research and Markets, 2015).

Bioplastics are now commonly used in the manufacture of bottles, cups, pots, bowls, flexible films, and other products. Packaging products from bioplastics are used in the packing of fresh food, dry snacks, candy, bakery goods, juice bottles, and meat trays as well as coatings for beverage cups, films, and card stock. Production capacity of bio-based plastic polymers is projected to triple from 5.1 million metric tons in 2013 (2% of total polymer production) to 17 million metric tons in 2020 (Bioplastics Magazine, 2015). Most bio-based plastics are made in large bio-refinery systems that produce various other products such as sweeteners and vegetable oil. Corn is currently the primary feedstock for producing bio-based plastics, with potatoes and other starch crops also used in lesser amounts (MarketWatch, 2016). The availability of cheap, bio-based raw materials has contributed greatly to the growing bio-materials market. It is likely that recent declines in crude oil prices will impact the bio-based materials market in the near term.

## The electronic waste challenge

Globally, electronic waste (e-waste) is one of the fastest-growing waste-streams. E-waste quantities expanded from 19.5 million tons in 1990 to 57.4 million tons in 2010, with future projected growth of 4–5% per year (Huisman, 2012). E-waste is different than most other waste-streams because of its individual product complexity and the amount of personally identifiable information present on used devices. Additionally, a variety of hazardous materials are commonly found in e-waste, including mercury, lead, and cadmium.

U.S. consumers currently purchase more than 1 billion electronic devices each year in the form of computers, monitors, televisions, and mobile phones. Additionally, Americans currently have about 3.8 billion devices stored in households – an average of 28 devices per household (Mars *et al.*, 2016, pp. 37-40). The production of electronic devices requires considerable investment of natural resources that are lost when e-waste is disposed of. A UN study found that the manufacturing of a computer and its screen takes at least 530 pounds of fossil fuels, 48 pounds of chemicals and 1.5 tons of water (Kuehr and Williams, 2003). E-waste also contains many valuable materials (e.g., iron, copper, aluminum, and plastics) and precious metals (e.g., gold, silver, platinum, and palladium) that can be recycled. The gold content from e-waste in 2014 was roughly 300 tons, which is equivalent to 11% of the global gold production from mines in 2013 (Baldé *et al.*, 2015, pp. 20-26).

Only 16% of the world's e-waste was recycled in 2014. E-waste recycling typically begins with collection by designated organizations, producers and/or by governments. Consumers with e-waste in need of recycling typically donate it to retailers, municipal collection points, and/or processing facilities. In the European Union, roughly 40% of e-waste is managed in this manner. In China and Japan, the e-waste recycling rate is around 24–30%. The U.S.A. and Canada recycle e-waste at a rate of about 12%, whereas Australia only recycles about 1%. Most of the items discarded as e-waste are actually not waste at all, but rather electronic devices that are still in working order and could potentially be marketed for reuse (Baldé *et al.*, 2015, pp. 20-26).

State-of-the-art processing facilities utilize sophisticated chemical processes to recover the valuable precious metals from e-waste but the metals are usually harvested at very low rates. For instance, printed wiring boards – present in many devices – contain precious specialty metals with high economic value such as gold, palladium, and silver. Current materials recovery processes used by e-waste recyclers recover gold at rates of 15–20% from printed wiring boards that are selectively removed from e-waste (Mars *et al.*, 2016, pp. 37-40).

Based on current e-waste disposal rates, Americans throw out phones containing over US$60 million in gold and silver each year. Recycling circuit boards can be more profitable than mining for ore. One ton of circuit boards is estimated to contain 40–800 times more gold than one metric ton of ore. A ton of circuit boards also contains 30–40 times more copper than can be recovered from a ton of ore. Electronics recycling is becoming less profitable as manufacturers try to save costs by using fewer rare minerals such as gold and copper in their devices, while the resale value of commodities recovered by recycling, such as steel and oil-based plastics, has declined sharply in recent years (Risen, 2016).

Given the large investment of natural resources required to manufacture electronic devices and the relatively low material recovery rates associated with recycling processes, refurbishment and resale has become increasingly popular in recent years. The resale value of refurbished devices is considerably higher than the material value of the components. Refurbished electronic items such as phones and computers can now be purchased at 50–75% of the price of a new product and frequently come with a warranty. Industry players estimate this market to be anywhere between US$15 billion and US$20 billion, in a fragmented sector comprised mostly of small regional players. With increasing demand for refurbished products, the segment is expected to grow exponentially in coming years, particularly if manufacturers incorporate features that make devices easier to disassemble and repair (Singhal, 2015).

## Energy efficiency

The worldwide mix of energy sources has changed a great deal over the past several decades. Coal and natural gas are still the most widely used fuels in electricity generation but there have been significant shifts regarding other energy sources. Power generation from nuclear power increased rapidly from the 1970s through the 1980s. The use of oil for power generation declined after the late 1970s, when sharp increases in oil prices encouraged power generators to substitute other energy sources for oil. However, oil has consistently provided the primary source of transportation fuel. According to the International Energy Agency (2016a) the global energy supply was provided by the following mix of resources in 2015:

- Oil 31.3%
- Coal 28.6%
- Natural gas 21.2%
- Renewables 14.1%
- Nuclear 4.8%

Each year, much of the energy the world consumes is wasted through heat loss, transmission loss, and inefficient technology. These losses squander billions of dollars for families and organizations and lead to unnecessary emissions and pollution. Energy efficiency is one of the easiest and most cost-effective ways to combat climate change, clean the air we breathe, improve the competitiveness of businesses, and reduce energy costs for consumers.

Much progress has been made in the past several decades regarding energy efficiency. The American Council for an Energy Efficient Economy (ACEEE) produces an International Energy Efficiency Scorecard that ranks the world's 23 largest energy-consuming economies on their energy efficiency policies and programs. Together these countries represent 75% of all the energy consumed on the planet and over 80% of the world's GDP. Based on the ACEEE's 2016 analysis regarding 35 categories of energy efficiency, Germany ranked first, followed by Japan and Italy, which tied for second. France and the U.K. rounded out the top five. The U.S.A. tied for eighth place with South Korea. The bottom five performers comprised Mexico, Thailand, South Africa, Brazil, and Saudi Arabia (ACEEE, 2017).

The International Energy Agency (2014, pp. 135-159) estimates that annual global investment in energy efficiency measures is currently about US$300 billion. The three regions that dominate global investment in energy efficiency are the European Union (27%), North America (20%), and China (20%). These investments are directed to a variety of transportation, industrial, commercial, and residential applications.

The U.S. Department of Energy's Better Building Alliance (Kwatra and Essig, 2014) noted that a typical commercial building can save 20% on its energy bills just by commissioning existing systems so that they operate as intended. They concluded that, on average, the projects they analyzed achieved a range of energy savings with favorable economics and achieved returns on investment (ROI) of 10–40%.

Alcorta *et al.* (2014) studied energy efficiency investments associated with 119 projects across nine manufacturing sectors in developing countries. They concluded that the investments were generally sound, with payback periods ranging from 0.9 to 2.9 years. Industrial operations can often generate considerable savings through upgrades to motors, pumps, compressors, lighting, process heating, and system controls. These upgrades produce additional benefits that extend beyond energy cost savings, including improvements in efficiency, productivity, product quality, and competitiveness. Energy efficiency measures can also improve the working environment for employees and facilitate compliance with environmental, health, and safety regulations. Opportunities for additional energy efficiency gains continue to be plentiful. An influential 2009 report from McKinsey analysts (McKinsey, 2009, pp. 1-14) projected that, by 2020, the U.S.A. could cut its annual energy use by 23% by investing US$520 billion in improvements such as sealing leaky building ducts and replacing inefficient household appliances with new, energy-saving models. The McKinsey analysis further projected that such measures would generate US$1.2 trillion in savings by 2020, and would reduce U.S. carbon emissions by up to one-sixth.

## Energy recovery and recycling

Historically, power grids have been developed by utilities to take advantage of economies of scale associated with large-scale power production and distribution. Utilities optimized their power plants and grids to produce and distribute as much electric power as possible because power was their only product. Unfortunately, this approach has resulted in power grids where much of the energy the world consumes is wasted through heat loss, transmission loss, and inefficient technology. For example, the USEPA estimates that the average efficiency of power plants driven by fossil fuels is about 33%. This means that two-thirds of the energy used to produce electricity at most power plants is wasted in the form of heat discharged to the atmosphere. Heat is valuable for multiple domestic and industrial applications but it dissipates quickly and must be utilized within close proximity to its production or it is lost.

Due to the sprawling nature of most power grids, it is not feasible to capture the waste heat and distribute it to distant customers that could potentially use the heat for productive purposes. Additionally, considerable energy is lost during the transmission and distribution (T&D) of the power as electricity travels over power lines. Within the five major power grids in the U.S.A., average T&D losses vary from 5.82% to 7.38%, with a national average of 6.18% (EPA, 2017c).

By producing and distributing electricity with smaller systems deployed on a more localized basis, the feasibility of capturing, distributing and utilizing the waste heat is greatly improved and the T&D losses can be minimized. Distributed generation systems can efficiently capture waste heat and distribute it to local facilities that can use it to heat a variety of assets such as buildings, industrial processes, and greenhouses. The recovered heat can also serve as an energy input for some chilling and refrigeration systems to produce chilled water, refrigeration, and air conditioning.

Distributed generation systems that recover and use waste heat are commonly referred to as "combined heat and power" (CHP) or "co-generation" plants. They normally convert and distribute 75–80% of the energy contained in the fuel to applications where it can be used productively. The most modern CHP plants can reach efficiencies of 90% or more (International Energy Agency, 2008, pp. 10-17). The economics of CHP systems can be quite favorable in applications that require considerable inputs of both power and heat.

CHP systems are most commonly powered by generators or turbines driven by combustible fuels. However, they can also be integrated into "microgrids" with other power sources, including geothermal, solar, wind, biofuels, and battery storage. According to the EPA, 88% of existing CHP plants are used for industrial purposes in energy-intensive applications. The other 12% are commercial and institutional entities such as hospitals, municipal and state governments, and public schools, colleges, and universities (U.S. Environmental Protection Agency, 2017).

Despite its efficiency and environmental benefits, CHP generation has languished at around 10% of global and U.S. capacity for more than a decade even though considerable opportunities exist for expanding deployment. Denmark obtains over half of its energy from CHP systems, which is a primary factor in its economy's ability to produce a unit of GDP with about 60% less energy than the U.S. economy. According to an analysis by the U.S. Department of Energy's Oak Ridge National Laboratory (Shipley *et al.*, 2008, pp. 19-28), CHP is "a proven and effective energy option, *deployable* in the short term, which can *help* address current and future U.S. energy needs." They projected that a large-scale expansion of CHP could provide 20% of U.S. generating capacity by 2030, generate US$234 billion in new investment, and create nearly 1 million jobs. Such an expansion would also reduce $CO_2$ emissions by more than 800 million tons per year, the equivalent of taking more than half the current U.S. passenger vehicles off the road. Unfortunately, industry experts frequently complain that electric utilities in most states utilize regulatory barriers to block CHP use by their customers.

## Renewable energy options

Since 1990, renewable energy sources have grown at an average annual rate of 2.2%, which is slightly higher than the total growth rate of the world energy supply (1.9%). World production of renewable energy grew 2.6% between 2013 and 2014 and currently

TABLE 5.1 **Composition of global renewable energy supply**

| Renewable energy source | Share of renewable energy supply (%) | |
|---|---|---|
| Biofuels (total) | | 72.8 |
| Solid biofuels and charcoal | 66.2 | |
| Liquid biofuels | 4.1 | |
| Biogases | 1.6 | |
| Renewable municipal waste | 0.8 | |
| Hydroelectric | | 17.7 |
| Geothermal | | 3.8 |
| Wind | | 3.3 |
| Solar and tidal | | 2.2 |

Source: International Energy Agency, 2016c.

comprises about 14% of the total primary energy supply (International Energy Agency, 2016c).

Due to its widespread noncommercial use (i.e., residential heating and cooking) in developing countries, solid biofuels (wood, charcoal, dung, etc.) are by far the largest renewable energy source, representing 10.4% of world energy consumption and 72.8% of global renewables supply. The second-largest source is hydroelectric power, which provides 2.5% of world energy supply and 17.7% of renewables. Geothermal, liquid biofuels, biogases, solar, wind, and tidal each hold a smaller share and make up the rest of the renewable energy supply (Table 5.1).

According to the International Energy Agency (2016c), since 1990, renewable electricity generation worldwide has grown on average by 3.6% per year, slightly above the total electricity generation growth rate of 2.9%. Renewables are now the second-largest source of global electricity production, accounting for 22.3% of world generation in 2014. Renewables come in behind coal (40.7%) and ahead of gas (21.6%), nuclear (10.6%), and oil (4.3%).

Hydroelectricity supplies the vast majority of renewable electricity (73.2%), which represents 16.4% of total world electricity generation. Following rapid recent growth, solar, wind, geothermal, and tide energies accounted for 4.2% of world electricity production in 2014 and 18.7% of total renewable electricity.

In the U.S.A., renewables supply about 13% of total electricity generation, ranking fourth behind natural gas (33%), coal (33%), and nuclear (20%) (EIA, 2015). The 13% estimate is virtually identical to the renewable electricity rates achieved in the 1930s, when large hydropower projects came online. The current composition of the U.S. renewable power mix is as follows:

- Hydroelectric: 46%
- Wind: 35%
- Biomass (wood): 8%
- Solar: 5%
- Biomass (waste): 3%
- Geothermal: 3%

Although progress in the implementation of renewable power generation has been steady, some sources of renewable power possess limitations that can restrict more rapid deployment.

Power demands tend to vary considerably throughout the day and from season to season. To ensure grid reliability for consumers, the power load must be balanced on a continuous basis. Generation technologies that can reliably adjust their output in response to changing demand are considered "**dispatchable**" and are more valuable to a grid system than less responsive, "**non-dispatchable**" technologies. Due to intermittency issues caused by fluctuations in weather and day length, wind, solar, and hydropower are not considered to be dispatchable. They cannot reliably respond to fluctuating power demands without some form of backup generation. Backup generation is usually provided through either:

- Conventional power generation with generators or turbines, or
- Energy storage (e.g., power stored in batteries or water stored in reservoirs)

In some applications, increased penetration from renewables – in place of conventional baseload generation – has not produced the expected reductions in fuel consumption and greenhouse gas emissions. Unfortunately, wind and solar tend to go offline quickly as a result of passing clouds and storms, and most conventional generation cannot be started quickly enough to fill the void. As wind and solar go offline, stored energy or conventional power generation has to be deployed rapidly to replace it in order to avoid power supply disruptions. To ensure the power supply is not interrupted, generators and turbines, powered by fossil fuels, are often kept "idling" at minimum load during the day even though the power they produce is frequently wasted because it is not needed (Budischak *et al.*, 2013). The intermittency of renewables has led to greater demand for the flexibility of gas-fired power plants in some markets.

Two factors commonly used to assess the value of power generation technologies are:

1. **The levelized cost of electricity:** This represents the per-kilowatt hour cost of building and operating a generating plant over an assumed financial life and duty cycle.
2. **Capacity factor:** This is the contribution that a given technology can make to the performance of the overall system. It is usually expressed as the percentage of power load that can reliably be served due to the addition of the power source.

**TABLE 5.2 U.S. national averages for total system electricity cost and capacity**
(Renewable energy sources are displayed in bold)

| Technology | Capacity factor (%) | Levelized cost of electricity (US$/MWh) |
|---|---|---|
| **Geothermal** | **91** | **42.3** |
| Conventional natural gas | 87 | 56.4 |
| **Wind (onshore)** | **42** | **58.5** |
| **Hydroelectric** | **60** | **63.7** |
| **Solar photovoltaic** | **26** | **74.2** |
| Conventional coal | 85 | 95.1 |
| Advanced nuclear | 90 | 99.7 |
| **Biomass** | **83** | **100.5** |

Note: Costs do not include subsidies.
Source: EIA, 2016c.

Table 5.2 summarizes the levelized cost and capacity factors for various sources of electrical power projected by the U.S. Energy Information Administration for 2018–22 (EIA, 2016c). As shown, the conventional power technologies (natural gas, coal, and nuclear) are all considered dispatchable, with capacity factors ranging from 85% to 90%. This is a major factor that contributes to the selection of these technologies by utilities because they can be counted on to address baseload power needs. The levelized costs (US$/MWh) for conventional technologies are estimated to be US$56.4 for natural gas, US$95.1 for conventional coal, and US$99.7 for conventional nuclear power.

The renewable energy technologies exhibit a wide range of capacity factors and levelized costs:

- Geothermal energy possesses both the lowest levelized cost (US$42.3/MWh) and the highest capacity factor (91%) of any power source.

- Hydroelectric power is projected to cost US$63.7/MWh but its 60% capacity factor, although considerably higher than wind and solar, prevents it from being considered dispatchable. However, it is noteworthy that hydroelectric power can be stored in reservoirs as a source of potential energy.
- Solar photovoltaic (PV) is projected to cost US$74.2/MWh but possesses the lowest capacity factor at 26%.
- Onshore wind power is considerably cheaper with a levelized cost of just US$58.5/MWh, but it also possesses a relatively low capacity factor of 42%.
- Biomass-based power generation can be considered dispatchable because of its relatively high capacity factor of 83%, but its levelized cost of US$100.5/MWh is projected to be the highest of all options.

Considerable investment continues to pour into wind and solar technology despite their low capacity factors and their inability to address baseload needs. Germany has added extraordinary solar and wind capacity in recent years but total power generation has not kept pace. It has achieved very low capacity factors, averaging just 17% for wind and 11% for solar (Budischak *et al.*, 2013).

The popularity of wind and solar has resulted from a variety of influences, such as marketing by the wind and solar industry, significant price declines for equipment, standardization of financing models by the financial sector, and lobbying by a variety of vested interests. Additionally, the highly visual nature of these technologies has made them recognized symbols of green technology. Various organizations have adopted images of wind and solar as symbols that provide visual depictions of their values and their commitment to protecting the environment.

Multiple opportunities exist to expand the deployment of biomass, hydroelectric, and geothermal energy sources. They all offer much higher capacity factors than wind and solar which makes them more dispatchable and reliable for meeting baseload needs. They are also cost competitive in many applications. Typical opportunities for expanding biomass, hydroelectric, and geothermal energy deployment include:

- **Biomass-based energy:** The first generation of liquid biofuels was produced primarily from food crops and created a great deal of controversy with respect to sacrificing food production to make fuel. However, new biofuels are emerging that are created from diverse sources of biomass, including municipal waste, waste tires, waste plastics, algae, and biomass produced on underutilized lands such as abandoned mines, and the strips of land set aside next to roadways or in medians.

  Biogas can be produced all over the world from diverse biomass sources, including livestock waste, food waste, and municipal waste-water. The wastes can be placed in a large vessel that serves as a reactor for anaerobic digestion by common microbes. The relatively simple process produces renewable biogas that can fuel conventional generators and turbines.

- **Hydroelectric power:** This offers the least expensive way to store potential energy (about one-fifth the cost of battery storage) through water collected in reservoirs. As long as the reservoirs contain sufficient water to produce the needed head pressure to drive the turbines, the stored energy can be released as needed to maintain grid reliability. Norway is currently the world's most prolific user of renewable energy thanks to its utilization of hydropower resources. More than half of Norway's domestic energy consumption is supplied by renewable hydropower (Science Daily, 2015). Environmental groups frequently object to new hydropower projects because the flooding created by reservoir construction can disrupt sensitive ecosystems. However, thousands of reservoirs already exist globally that could potentially be retrofitted to incorporate hydropower, and many organizations are in the process of evaluating and prioritizing these opportunities.

- **Geothermal power:** Given that geothermal power generation possesses both the highest capacity factor and lowest levelized cost of *any* power source, its potential for adding reliable baseload power has barely been tapped. Geothermal energy uses heat stored in the Earth's core to generate power and to heat and cool buildings. Unlike wind and solar, this renewable energy resource is available 24 hours a day, 365 days a year, making it the most dependable resource for meeting baseload power needs. Historically, geothermal power development has been restricted to regions where geothermal resources exist close to the surface. This restriction has resulted from the fact that the biggest cost associated with geothermal power development is usually drilling. However, geothermal resources can actually be developed anywhere, provided that the developer is willing to drill deep enough to access the resource. Advances in drilling technology and the potential for using existing wells that were drilled for other purposes (e.g., oil and gas production) could establish a breakthrough for this underutilized technology.

Energy storage is a rapidly growing market that can potentially improve the capacity factors for wind and solar energy. However, energy storage is still in its infancy because it tends to be too expensive for widespread deployment. Recent advances in battery technology for hybrid and electric vehicles have been encouraging but significant subsidies will be required in the near term if battery technology is expected to play a significant role in storing power generated by wind and solar. The cost of rechargeable batteries for energy storage needs to decline drastically for the storage battery market to gain real momentum. An analysis by Citi (2015, pp. 74-81) estimated that battery storage of grid power will need to achieve a price point of US$230/KWh to be economically feasible. Some Japanese electronics makers currently sell storage batteries for residential use, but the products cost more than US$1,000/KWh. For comparison purposes, electric vehicle battery prices cost about US$1,000/KWh in 2010, when the first commercial electric vehicles appeared. With market growth in the electric vehicle sector, the cost is now around US$500/KWh, and with volume growth and improving cost of goods sold, Citi projected future annual cost declines will occur at a rate of about 10%.

# 6

# Charting a prosperous path forward

> The pessimist complains about the wind; the optimist expects it to change; the realist adjusts the sails.
>
> William Arthur Ward

Forward-thinking leaders approach constraint headwinds proactively and recognize them as opportunities for increasing effectiveness. Instead of focusing on the external factors that contribute to constraints, they focus internally on their own practices, processes, products, and systems to develop measures that can address the root causes. Some organizations have actually been able to transform their organizations so significantly that the headwinds have become a source of tailwinds that improve effectiveness and competitive advantage. To accomplish this type of transformation requires careful analysis of the systems that comprise the organization and the ways in which various aspects of the systems contribute to constraints.

## Lessons from an unsustainable past

> O, wind, if winter comes can spring be far behind?
> Percy Bysshe Shelley

The growth of constraint headwinds is the result of converging trends that continue to grow in both magnitude and intensity. Increases in population and industrialization have led to corresponding increases in resource demand, regulations, and uncertainty. As these trends progress, they generate negative consequences, such as declines in resource availability, resource quality, quality of life, and economic prosperity. These factors combine to produce a formidable mix of constraints that, in turn, create strong headwinds with respect to environmental protection, community development, and economic growth.

Changes happened so rapidly during early industrialization that society simply didn't anticipate the volume, magnitude, and complexity of the resulting side-effects and impacts. Most of the corrective measures that were implemented were a reaction to problems after they had already escalated to crisis level. In some instances, progress has been made to proactively address headwinds through measures that prevent waste and hazards; improve the life-cycle of processes and products; and optimize systems to create a circular flow of energy and materials. However, these gains have historically lagged behind the growing constraints and impacts, resulting in considerable damage to community and environmental assets and stagnant economic growth. If the consequences had been anticipated in advance, measures could have been undertaken earlier to either prevent them from happening or mitigate them before the impacts became severe. Future strategies need to focus on tackling the root causes of these problems through the development and implementation of better processes, products, and systems.

The trends that create constraint headwinds have been around for many years and they continue to grow with considerable momentum. On the surface, these trends appear to be regional or even global in nature – the result of factors that are perceived to be outside the influence of most organizations. For example, a company dependent on resources that are declining in availability and quality (minerals, ore, electricity, water, landfill space, infrastructure, skilled labor, etc.) may experience escalating costs, supply chain disruptions, and production delays. Regulators may intervene and modify policies to control and reduce their utilization of resources. The escalating volume and complexity of the constraints can be overwhelming to those considering how to manage them. They often choose to scale back operations until conditions improve or they lobby government officials for more favorable policies. If the conditions persist or escalate, they may choose to shut down the constrained operations and relocate to another region with fewer constraints and more favorable policies. If the company disregards the constraints and maintains its production levels, the availability and quality of resources will be further constrained and the headwinds will escalate.

Organizations can sometimes utilize wasteful or unsafe practices to produce short-term financial gains, but these benefits cannot be sustained in the long term. Many organizations continue to operate processes and make products with a variety of deficiencies that limit their safety, productivity, efficiency, and quality. These deficiencies ultimately result in accidents, wasted resources, emissions, and pollution. Decisions to manage organizations in this way are negligent and unethical. Organizations that depend on profiting from practices or products that create more harm than good are vulnerable in today's age of transparency and accountability. Even the best public relations and marketing campaigns will not preserve an organization's reputation and license to operate. If a business makes people unhealthy, or depends on exploiting workers, or can be tied to environmental degradation, it risks mass customer migration to businesses with better performance.

Diverse interests, including governments, advocacy groups, and forward-thinking businesses, now recognize that their long-term success is dependent on more than just short-term financial performance. Equally important are the quality of life in the communities they interact with and the quality of the environment they depend on for resources and services. The relationships between the quality of the natural environment, the quality of communities, and economic development are much better understood today than in the early stages of industrialization. Practices that exploit environmental and/or community resources to achieve financial gains are never viable in the long term. Declines in the quality and availability of environmental and community assets eventually impact the organizations that depend on them for important resources.

## The importance of balance

> You can find peace amidst the storms that threaten you.
>
> Joseph B. Wirthlin

As man-made economic and social systems have matured, so has our understanding of the interdependent relationships between the economy, society, and the environment. Practices that improve one of these three dimensions at the expense of the others are simply not sustainable in the long run. If we engage in practices that deplete the environment for the sake of economic gain, the environmental resources will eventually be depleted, revenue will decline along with job availability and the tax base, and communities will suffer. Likewise, if we decimate business activity for the sake of protecting the environment, unemployment will increase and the tax base will decline, reducing the financial resources needed for environmental protection and community development.

## Identifying and prioritizing opportunities

Sustainability is a topic of high interest these days among very diverse interests. Countless organizations and their leaders are searching to determine:

- How do sustainability challenges constrain their current and future effectiveness?
- What are their potential roles and capabilities for addressing the challenges?
- What opportunities exist to prosper from addressing sustainability challenges?

Before an organization can develop a sound strategy for improving sustainability performance, it needs to identify how its existing systems contribute to constraint headwinds at a root cause level. Once the root causes have been identified, organizations can develop and implement measures based on sound sustainability principles that will defeat the root causes of the constraint headwinds. Implementing measures that address the root causes of constraint headwinds can produce a variety of economic, social, and environmental benefits. Such measures can serve as a compass for guiding the organization's strategy toward "doing the right things" to improve sustainability performance.

Leading organizations all over the world are taking proactive measures to address constraint headwinds by incorporating tactics into their strategy that preserve and improve the quality of community and environmental assets. Such measures are crucial for safeguarding their long-term survival and effectiveness. The key to future prosperity is finding and pursuing opportunities that generate lucrative revenue while also benefiting communities and the environment. Fortunately, such opportunities are plentiful and will continue to be so for the foreseeable future.

# 7

# The rationale for sustainable strategy

> Thought is the wind, knowledge the sail, and mankind the vessel.
>
> Augustus Hare

The UN and countless other organizations talk continually about the need to address the growing constraints affecting the planet, and urge society to take steps that will end extreme poverty, fight inequality and injustice, and improve environmental health. The UN defines the term "sustainability" as *meeting the needs of the present without compromising the ability of future generations to meet their own needs*. This is a noble concept and one that we all need to take seriously for the sake of our children and grandchildren. The UN has also established a list of 17 sustainable development goals for 2030 that it believes are critical for "meeting citizens' aspirations for peace, prosperity, and wellbeing, and to preserve our planet" (UN Development Programme, 2016). Definitions and goals such as those advocated by the UN are helpful for establishing a vision of what a more sustainable society could look like. They can also establish some meaningful milestones with respect to what it will take to accomplish this vision. However, definitions and goals do not provide any actionable guidance with respect to how to actually go about improving an organization's sustainability performance.

A new phase of industrialization is underway that rewards organizations who operate more sustainably and punishes those who do not adapt to changing conditions and demands. Organizations that have historically benefited from processes and products that are wasteful now face considerable constraint headwinds. This includes businesses that consume large amounts of resources or cause wasteful negative impacts such as pollution or hazards to communities, people, and/or the environment. Such businesses are not likely to endure in the future.

To ensure the long-term sustainability of an enterprise, a community, a nation, and a planet, organizations need to focus on balancing and improving performance regarding three dimensions:

1. Economic growth (revenue, costs, innovation, and risk)
2. Social responsibility (employees and community stakeholders)
3. Environmental stewardship (air, water, land, materials, and energy)

Many leaders now recognize that the three dimensions are inextricably linked and, to achieve long-term prosperity, they must be managed together to achieve benefits for all three. It is no coincidence that regions with strong economic performance also tend to be better stewards of their communities and the environment. Organizations all over the world are pursuing strategies that simultaneously advance economic growth, community development, and environmental protection. Incorporating sustainability principles into organizational culture is not just about saving the planet or feeling good, it is key indicator of operational and management quality.

## The benefits of more sustainable performance

Multiple benefits can be achieved by operating more sustainably for diverse stakeholders, including employees, customers, dealers, communities, and the environment. Such benefits can include:

- **Environmental stewardship benefits**
  - Reduced emissions, by-products, and wastes
  - Improved regulatory compliance
  - Preservation of resources for future generations
- **Social responsibility benefits**
  - Reduced hazards and enhanced safety conditions
  - Better quality of life in communities
  - Improved opportunities for employees
- **Economic development benefits**
  - Reduced resource use and costs
  - Improved value to customers
  - Improved reputation and market differentiation

Sustainability performance is improved when productivity, effectiveness, and prosperity are accomplished through measures that ensure future capabilities are preserved; and environmental and community resources are likewise protected or improved.

## Elements of effective sustainability programs

To ensure their long-term viability, organizations are now compelled to demonstrate a visible and authentic commitment to sustainability. Key elements that are common to effective sustainability programs include:

- **Organizational leadership:** Leaders are very engaged and develop well-articulated strategy and focus areas that are woven into the overall enterprise strategy and values as opposed to establishing a separate sustainability strategy.
- **Compelling business case:** Leaders work with their employees and other stakeholders to develop proactive options for enabling sustainable progress that often result in compelling business prospects.
- **Effective communications:** The importance and benefits of operating more sustainably are well communicated and understood by employees and other stakeholders.
- **Employee engagement:** Leading organizations support creativity, innovation, and risk taking by encouraging open dialogue across the organization to explore and suggest opportunities to develop and implement more sustainable practices.
- **Stakeholder engagement:** Leading organizations pursue opportunities in the organization's operations as well as opportunities to engage with investors, suppliers, customers, community leaders, and environmental advocates.
- **Reporting, goals, and metrics:** Leading organizations work with stakeholders to identify key aspects that are most material, set aggressive goals, and regularly measure and

report their progress. They build awareness of sustainability challenges and programs both within the company and among stakeholders.

- **Formulate and embed implementation strategy:** A strategy for achieving sustainability goals with a clear business case is established and integrated into the organization's overall strategy.

These characteristics can be used to help develop a vision of what success looks like and a list of tasks to undertake. They describe what it means to "do things right" but they don't describe what it means to "do the right things." More directional guidance is needed to steer an organization's culture, strategy, processes, products, systems, innovations, and designs toward more sustainable performance. Directional guidance for improving sustainability performance is provided in Chapters 9–13.

## Drivers for improving sustainability

Drivers for improving an organization's sustainability performance come from a variety of internal and external sources. Some of these drivers are business focused and address constraint headwinds directly while also generating a combination of economic, environmental, and social benefits. Strategies guided by these drivers can produce compelling business opportunities that greatly improve the organization's effectiveness. Other drivers come from external sources such as government agencies, nongovernmental organizations (NGOs), and investors. These drivers can also lead to considerable benefits but frequently come with a heavy bureaucratic burden. A summary of the various external and business drivers is provided in Table 7.1, followed by more detailed descriptions.

TABLE 7.1 **Business and external drivers**

| Business drivers | External drivers |
|---|---|
| • Recruiting and retention<br>• Operational excellence<br>• Financial performance<br>• Business growth<br>• Reputation and brand<br>• Shareholder value<br>• Innovation | • Transparency<br>• Reporting<br>• Metrics<br>• Ratings and certifications<br>• Compliance<br>• Public relations |

## External drivers for improving sustainability

A wide range of external stakeholder interests – comprised of NGOs, ratings organizations, investors, government agencies, and customers – are now pushing organizations to provide ever-increasing information and transparency regarding their sustainability performance. They believe that additional information and transparency will empower decision-makers to take action toward a more sustainable economy and world. Many investors care about sustainability performance because of their focus on efficiency, risk, and growth. They recognize that if a company can produce a product that meets customer demands while also providing social and environmental benefits, the company may have strong future prospects. However, investors tend to make decisions based on data while companies tend to use words to describe their sustainability efforts. Translating between the two is critical for aligning the interests and strategies of both entities. More detailed summaries of the external drivers for improving sustainability performance are provided below.

## Reporting and transparency

Access to information continues to grow exponentially around the world and efforts to keep information secret are becoming increasingly futile. Stakeholders are increasingly requesting information regarding organizations' sustainability performance. Many organizations are responding to these trends by becoming more proactive with their information sharing and transparency initiatives. Thousands of organizations, including 92% of the world's 250 largest corporations, currently report on their sustainability performance (GRI, 2016a, pp. 7-15). Winston (2014, p. 52) concluded that:

> Radical transparency is here to stay. Every company that wants to remain competitive needs to answer tough questions. Many companies talk about how their environmental or social performance is often a tiebreaker in sales: the company and products with a better backstory, supported by good data, land more contracts.

Sustainability reporting can help organizations measure, assess, understand, and communicate their economic, environmental, social, and governance performance. Organizations can use this information to set meaningful goals, establish priorities, and manage change more effectively. The value of the sustainability reporting process is that it ensures that organizations consider their impacts on important issues and it provides a venue for them to be transparent about the risks and opportunities they face. Many organizations engage stakeholders to provide critical input in identifying their most critical risks and opportunities, particularly those that are nonfinancial. The increased transparency provided by sustainability reporting can lead to better decision-making, which helps build and maintain trust in businesses and governments. Providing transparency, with regard to sustainability performance, often creates considerable incentives for improving performance because anyone can assess, rank, and critique the organization's progress.

A growing number of investors use sustainability reports as a resource to assess if the organization's sustainability efforts are focused on the material issues that affect its ability to survive and prosper. Investors also want to know the business specifics of how sustainability is creating value for organizations they consider for investment. Multiple factors may impact an investor's assessment of how sustainability is contributing to an organization's value, including cost of capital, levels of innovation, and market impacts. The Global Reporting Initiative (GRI) is an international standards organization focused on sustainability. With thousands of reporting organizations in over 90 countries, it provides the world's most widely used standards on sustainability reporting and disclosure. The GRI has pioneered sustainability reporting since the late 1990s, transforming it from a niche practice into one now adopted by a growing majority of organizations. The GRI reporting format includes 79 separate performance indicators in its "Standards for Sustainability Reporting" (GRI, 2016b).

In recent years, some groups have begun advocating for more comprehensive reports that combine the analysis of financial and nonfinancial performance into a single "integrated" report. In the U.S.A., the Sustainability Accounting Standards Board (SASB) is working to develop rules governing public disclosure by operating within the current system of financial regulation. The SASB's goal is to integrate its standards into Form 10-K, the statutory financial report which public companies must file each year with the U.S. Securities and Exchange Commission. The SASB plans to establish industry-specific reporting standards that will facilitate comparison and benchmarking, and has devised a Sustainable Industry Classification System covering ten sectors and more than 80 industries (SASB, 2016).

The European Commission (2016) manages rules on information prepared and disclosed by EU companies, including financial statements and non-financial information. Large public-interest entities (listed companies, banks, insurance undertakings, and other companies that are so designated by member states) with more than 500 employees should disclose in their management report relevant and useful information on their policies, main risks, and outcomes relating to at least:

- Environmental matters
- Social and employee aspects
- Respect for human rights
- Anticorruption and bribery issues
- Diversity in their board of directors

There is significant flexibility for companies to disclose relevant information (including reporting in a separate report), as well as whether they rely on international, European or national guidelines.

## Sustainability ratings

Dozens of external ratings, rankings, indices, and awards currently exist that seek to measure and assess corporate sustainability performance. These programs typically involve questionnaires that solicit hundreds of data inputs across multiple aspects of the organization's performance. Some organizations rely on such ratings to gauge and validate their own sustainability efforts, with some even linking management performance evaluation and compensation to the external ratings. Certain stakeholders (e.g., investors, consumers, and employees) utilize these ratings to help inform their decisions regarding investment, purchasing, employment, and so on.

Therefore, for the ratings to be used effectively, they must be robust, accurate, and credible. The best-known ratings programs include:

- Dow Jones Sustainability Index
- Carbon Disclosure Project (CDP)
- FTSE4Good Index Series
- Fortune's Most Admired Companies
- The Global 100 Most Sustainable Corporations in the World
- Bloomberg Environmental and Social Governance Data

The CDP and Dow Jones Sustainability Index are commonly regarded as the most credible by sustainability professionals (Sadowsky, 2014).

Considerable irony exists with respect to the extraordinary emphasis on these ratings programs. To be considered a *leader* in your sector, you have to *follow* the prescription of NGO experts who have little expertise in that sector. For example, while serving as the Global Director of Sustainable Development for Caterpillar, I interacted with a "sector expert" for one of the leading rating programs whose previous experience and training was focused on tending bar. It is also noteworthy that many of the ratings programs offer reporter/consulting services to organizations interested in improving their scores.

## Keeping score versus scoring

Preparing credible sustainability reports that meet external standards and compiling data for ratings surveys are exhaustive tasks. For large organizations, the processes require months to complete and hundreds of hours of input from dozens of individuals in diverse roles. The rationale behind the relentless emphasis on measurement and reporting is rooted in the long-recognized relationship between measurement and management. The importance of this relationship has been emphasized by multiple business experts such as Peter Drucker, W. Edwards Deming, and Tom Peters. It also has roots in science and mathematics, including the early-20th-century Irish physicist Lord Kelvin, and the 16th-century Austrian mathematician and compass inventor, Rheticus (Henderson, 2015). Whatever the original source, the relationship between measurement and management has produced some of the most frequently used clichés in all of business. Here are a few examples (authors are not provided because consensus regarding authorship is impossible to validate):

- "What gets measured gets managed"
- "What doesn't get measured doesn't get managed"
- "You can't manage what you don't measure"
- "You manage what you measure"
- "What gets measured gets done"
- "To measure is to know"
- "If you can measure it, you can manage it"
- "What's measured improves"

Obviously, the relationship between measurement and management is important. Measuring the right things can produce valuable information needed to set goals and priorities, and assess progress. Measurement can also help incentivize improvement because people know that their performance is being monitored and they will be held accountable for achieving results. Every manager has heard employees proclaim that "if it's not in my performance goals and objectives, I can't make it a priority." The scorekeeping aspect of measurements can appeal to the competitive nature of some people. Without a measure, there is no way to determine whether you have won and, therefore, employees have less motivation to get something done. However, doing a good job of measuring does not ensure that progress will be achieved.

## Clichés don't produce results

Unfortunately, the extraordinary emphasis on the relationship between measurement and management has led many individuals and organizations to conclude that a strong degree of causation exists between the two factors. They assume that increases in measurement and reporting will provide the stimulus necessary to drive changes that will lead to performance improvements. This notion has become an important aspect of thought leadership for many sustainability experts who strongly believe that catalyzing improvement will be best achieved through more, and better, measurement and reporting. The logic behind this assumption is questionable given that cause/effect relationships between measurement and performance have been impossible to establish despite multiple attempts to do so in diverse applications. For instance:

- Does weighing yourself ensure you will lose weight?
- Does testing students more frequently improve their education?
- Does tracking quarterly revenue ensure that sales will increase?

Without question, measurements and reports are important management tools, but it is not appropriate to assume that they will serve as powerful drivers for change. The likelihood that measurement and reporting will significantly impact sustainability performance is doubtful unless, of course, sustainability is somehow different than other issues. A reasonable question to test this hypothesis might be phrased something like this:

- Does increasing sustainability measurement and reporting improve performance?

The answer to this question is NO – based on data published in the 2015 Carbon Disclosure Project (CDP) *Global Climate Change Report* (CDP, 2016, pp. 6-8). The report noted that 44% of companies responding to the CDP's most recent questionnaire have now set greenhouse gas reduction targets, compared with just 27% in 2010. However, these additional commitments to goals and reporting did not stop emissions growth. The report concluded that "On a like-for-like basis, direct ('Scope 1') emissions from the companies analyzed for this report grew 7% between 2010 and 2015. [Indirect] Scope 2 emissions, associated with purchased electricity, grew 11%."

To improve sustainability performance, change has to happen across a variety of aspects, such as strategy, planning, methods, designs, processes, products, and systems. Some progress can be achieved through incremental improvements but for the types of step change needed to overcome today's constraint headwinds, considerable innovation is necessary. Simply doing a better job of measuring and reporting impacts does not ensure that improvements will happen. These points seem rational to most individuals but they don't explain the overwhelming popularity and emphasis on reporting and ratings programs. Some of the various factors that contribute to this phenomenon are:

- **Compliance mindset:** An organization's sustainability programs are often managed by individuals who regularly work on prescriptive compliance with various environmental, health, and safety requirements (e.g., environmental, health, and safety professionals). Gathering data and reporting to authorities is well within their comfort zone because they perform these duties routinely to ensure compliance with various regulatory standards. Consequently, adding sustainability reporting to their responsibilities is compatible with their normal approach to work.

- **Innovation risk:** Leaders in all types of organization recognize the need for innovation in today's rapidly changing world. However, innovation requires risk and many leaders and organizations tend to be risk averse. See Chapter 14 for further discussion of this aspect.
- **Leadership advocacy:** To implement significant change in organizations, considerable leadership is required from effective champions who possess enough influence to overcome resistance to change. Individuals involved primarily in prescriptive reporting and compliance functions are not usually tasked in the role of leading organizational change. Consequently, effective sustainability champions must be recruited and cultivated broadly from diverse aspects of organizations. This can be very challenging.
- **Market access:** For consultants and NGOs interested in penetrating the sustainability market, business models and services focused on measurement and reporting are easier to establish than approaches focused on driving substantive change. Consequently, thousands of organizations have now entered the sustainability consulting market, a market that relentlessly emphasizes measurement and reporting. The frequency and consistency of their messaging continuously reinforces the perceived need for additional measurement and reporting.

In spite of the emphasis on measurement and reporting, leaders from some organizations are now questioning the value that escalating demands for measurement and reporting provides to stakeholders. Many have suggested that the resources invested in exhaustive data-gathering and reporting could be put to more productive use in activities that will actually drive improvements and produce results. Additionally, the prescriptive nature of the various surveys and reports is often perceived to be "just another set of standards to comply with" by the many individuals who are involved in obtaining and compiling the information. As this perspective grows, it creates challenges for sustainability advocates who are trying to convince their leadership that sustainability is good for business. It is difficult to convince top management that exhaustive measurement and reporting will improve the bottom line.

## Putting points on the board

When you attend an athletic event or other type of competitive contest, do you invest much time watching the scorekeepers, referees, or reporters? Probably not – unless they blow a call that negatively impacts the participant(s) you are rooting for, most spectators don't pay much attention to these individuals. We all recognize that they make important contributions toward ensuring the success of a competitive event. Without good scorekeeping, we wouldn't know who won the contest. Without good officiating, the rules wouldn't be followed and people would cheat. Good reporting provides helpful analysis of what went well, what didn't, and who made major contributions. While these roles are all important for ensuring an event's quality and success, if done well they should not directly affect the outcome. Contest outcomes should be determined by the innovative ways that participants put points on the board – or prevent opponents from doing likewise. In other words, scoring is far more important than keeping score. When it comes to sustainability performance, scoring is accomplished predominantly through innovations in strategy, planning, methods, designs, processes, products, and systems. Therefore, the key to driving sustainability performance is the implementation of more sustainable innovations.

## Compliance is NOT a strategy

Compliance with all applicable regulatory requirements is always a minimum expectation for ethical organizations. In fact, regulatory compliance is at the core of many organizations' strategies for engagement with communities and the environment. However, regulatory compliance alone is no longer adequate for ensuring that an organization's license to operate will be maintained in the long term. Regulations emerge when market-based systems fail to protect people, communities, the environment, and/or the economy from problems such as hazards, pollution, and unfair practices. **If organizations focus only on meeting requirements imposed by government agencies, they may miss out on more effective possibilities that can be achieved through innovation.** Taking reactive measures to address regulatory requirements can achieve compliance but such measures can be inefficient because they usually do not correct problems at their source.

The root causes of most problems are embedded in organizations' processes, products, and systems. Regulators do not usually possess the detailed information and technical knowledge required to analyze and address the root causes of all the problems they encounter. Instead of addressing problems at their source, regulators usually focus on measures that address problems after their consequences have escalated and surfaced to impact human, community, and environmental resources. Therefore, organizations with environmental and community engagement strategies based primarily on complying with regulatory requirements frequently miss out on opportunities to improve effectiveness. The following case study provides a comparison of the effectiveness of reactive compliance measures versus proactive measures that address problems at their sources.

**Case study: Addressing air quality at a machining operation**

I once worked with a machining company that operated a facility with dust levels that exceeded safety limits for extended worker exposure. Workers were required to wear uncomfortable dust masks to protect them from the dust. This measure protected the workers and was compliant with safety standards but it did not address the root cause of the problem. An analysis of the process that generated the dust revealed that the dust was actually raw material that was intended to be blended into a product. Unfortunately, antiquated grinding and mixing equipment was dispersing an unacceptable quantity of the valuable material into the air. Research into new grinding and mixing systems indicated that newer versions possessed features that could effectively direct the pulverized material into the finished product where it belonged instead of dispersing it to the air. This improvement could achieve more efficient use of the raw material, eliminate the dust problem, and create a healthier work environment, thereby eliminating the need for employees to wear uncomfortable dust masks. A very attractive return on investment was achieved by updating the equipment and improving material efficiency and product yield. Both measures (the dust masks and the machine modification) could achieve compliance and protect workers from the dust problem. However, the equipment modification was a far more effective solution because it prevented the problem at its source, achieving compliance with the safety standard while also improving efficiency and productivity.

The leaders of many organizations remain convinced that adding focus to environmental and social performance will erode their competitiveness. They believe it will add to costs and will not deliver immediate financial benefits. This outdated idea is rooted in the longstanding belief that the environment and the economy are at odds with each other. Several years ago, I witnessed a senior executive of a Fortune 100 company proclaim to his entire staff that he didn't subscribe to the organization's recent emphasis on improving sustainability culture and performance. He chuckled openly as he explained that he didn't really understand what it meant and questioned if he would still be able to drink out of a plastic water bottle. He further explained that, in his opinion, taking measures beyond what is required by law is a waste of resources and he scoffed at recommendations for more proactive initiatives. His views were reflective of someone who had spent the previous 40 years working under the assumption that business and the environment are at odds with each other. He was comfortable in his role of reacting to government regulation with the least possible response and believed that this approach was in his employer's best interest. From his perspective, sustainability issues were always liabilities that should be minimized as opposed to potential assets that could be maximized to improve competitive advantage. He completely missed the opportunity at hand. The end result was that he stifled considerable employee interest in improving the sustainability of the enterprise, and the company missed out on countless opportunities to improve effectiveness. Within two years of his remarks, the company's shareholder value had plummeted over 40%. Coincidence? Continue reading ... then decide.

Empowering regulators to guide sustainability strategy is comparable to allowing them to guide other core functions such as production, maintenance, purchasing, or quality assurance. Countless opportunities often exist to achieve compliance through preventative measures that address problems at their source. Unfortunately, these opportunities are frequently overlooked because, instead of proactively investigating and addressing root causes, businesses focus on reacting to the demands of regulators.

## Business drivers for improving sustainability

The quest for more sustainable practices and outcomes is now creating some of the most effective drivers for improving enterprise performance across multiple dimensions. Most leaders now agree that managing sustainability performance is no longer an option for organizations interested in thriving for the long term. A 2016 survey conducted by Accenture collected input from over 1,000 global CEOs representing 27 industries across 103 countries, and found that 97% believe that sustainability is important to the future success of their business. Further, 85% of the CEOs indicated they have embedded sustainability into the business even where they cannot quantify the benefits (Accenture Strategy, 2016). A 2014 McKinsey survey of over 1,000 CEOs revealed that 43% of CEOs say their companies seek to align sustainability with their overall business goals, mission, or values – up from 30% who said so in 2012 (McKinsey, 2014).

Here are some examples of how sustainability performance is linked to various aspects of organizational effectiveness and competitiveness. These drivers tend to be compatible with the priorities and culture of most forward-thinking organizations.

## Sustainability can drive recruiting and retention

Employees are the backbone of virtually all organizations. They provide the vision, creativity, and hard work required for productivity and leadership. A 2014 study by the Society for Human Resource Management (2014) indicates that 72% of organizations practice sustainable workplace initiatives and 50% of those organizations have a formal sustainable workplace policy that integrates these initiatives into their strategic planning process. The survey findings also reported additional human resource benefits from engaging in more sustainable practices, including attracting top talent (51%), improving employee retention (40%), and developing leadership (36%). It also found that the higher employees rate their organization's corporate citizenship, the more committed they are to the organization.

## Sustainability can drive operational excellence

Preventing defects and problems is a central theme of quality management systems. This approach can easily be extended to apply more broadly to sustainability aspects. Huge sustainability benefits can be achieved by improving the quality, safety, efficiency, and productivity of processes and products through measures that prevent and eliminate all types of risk and waste. Maximizing the life-cycle benefits of products and services – while minimizing the economic, social, and environmental costs of ownership – improves competitiveness by providing the utmost value for customers. Unruh *et al.* (2016) performed an analysis of more than 200 sustainability studies and reports, and concluded that nearly 90% of the documents indicated that solid sustainability practices drive improvements in operational performance.

## Sustainability can drive financial performance

Improved safety, efficiency, productivity, and quality performance inevitably leads to reduced costs, increased revenue, and higher profits. Embedding sustainability principles in all aspects of an enterprise can uncover opportunities for preventing wastes and hazards and reducing costs. Proactively addressing these opportunities can reduce the need to purchase resources such as materials, energy, water, and land, and can prevent costly accidents. A 2016 analysis performed by MIT Sloan collected input from 579 investors regarding the importance and impacts of various sustainability practices (Unruh *et al.*, 2016). Almost 75% of the respondents indicated that they "feel strongly that increased operational efficiency often accompanies sustainability progress." In addition, more than 80% of investor respondents indicated that good sustainability performance increases a company's potential for long-term value creation.

## Sustainability can drive business growth

The communities of the future – along with the infrastructure and organizations that support them – will all be designed, constructed, and managed using more sustainable technology. Growth opportunities associated with alternative energy sources, materials, methods, machines, sensors, and communications will be developed and integrated into systems that enable safer and more productive uses of all resources. A recent report from the Conference Board (Singer, 2015) studied revenue from portfolios of "more sustainable" products and services, defined by 12 S&P 100 companies between 2010 and 2013. Revenues from the "more sustainable" products and services grew by 91% – six times the rate of overall company results.

## Sustainability can drive reputation and brand

The better a company protects its reputation and builds brand trust, the more successful it will be at gaining and maintaining competitive differentiation. A 2014 Conference Board study (Singer, 2014) revealed that 88% of surveyed investment professionals, purchasing managers, and graduating university students believe that sustainability reputation is either "extremely" or "somewhat" important in the decisions they make to invest in, partner with, or work for a company.

## Sustainability can drive shareholder value

Growth, efficiency, and risk are core considerations for investors and sustainability is a critical component of each. Strong sustainability performance has been shown to be a solid indicator of a company's value, and investors are factoring companies' sustainability performance into their decisions to invest. A 2014 report from the International Finance Corporation (2014) concluded that companies listed on the Dow Jones Sustainability Index (companies that score in the top 10% based on a detailed assessment of the social, environmental, and economic performance of 2,500 companies) performed on average 36.1% better than the traditional Dow Jones Index over a period of five years. Of the 579 investors surveyed by MIT Sloan, 75% indicated that a company's good sustainability performance is materially important to their firms when making investment decisions (Unruh *et al.*, 2016). Likewise, poor sustainability performance can have a dramatic negative effect on a company's value: 44% of the respondents indicated that poor sustainability performance is a deal breaker – they won't invest in a company with poor sustainability performance. According to Arabella Advisors (2016), a firm that specializes in impact investing, 688 institutions and 58,399 individuals across 76 countries have committed to divest a combined US$5 trillion from fossil fuel companies to date. The level of divestment doubled over a 15-month period.

Recent incidents associated with Volkswagen, BP, and Lumber Liquidators are noteworthy examples of how negative sustainability incidents can decimate shareholder value:

- **Volkswagen emissions scandal:** In September of 2015, the German auto-maker lost nearly one-third of its market value in two days when the U.S. EPA accused Volkswagen of using software that could cheat emissions tests associated with diesel engines they installed in 11 million cars. Several Volkswagen suppliers and customers also suffered large declines in market value. The impacts of the scandal even extended to competitors Ford, General Motors, and Fiat-Chrysler as the value of their shares declined 2–4% over the same two-day period (Hadi, 2015).

- **BP Deepwater Horizon explosion:** On April 20, 2010, while drilling for oil in the Gulf of Mexico, the drill encountered high-pressure natural gas that caused a blowout in the piping. Natural gas was released to the rig's main deck where it exploded, killing 11 crew members and releasing nearly 5 million barrels of oil. About 68,000 square miles of ocean was impacted along with hundreds of miles of coastline in Louisiana, Mississippi, Florida, and Alabama. Investigations revealed that the blowout resulted from a combination of defective materials and failures in the Deepwater Horizon's safety systems. From April 19 to June 25, 2010, BP's share price declined by 55% (Chamberlin, 2014). In

June 2016, a U.S. Federal Judge approved a US$20.8 billion settlement for Clean Water Act penalties and other environmental damage (Barrett, 2015).

- **Lumber Liquidators formaldehyde scandal:** Following the March 2015 airing of a *60 Minutes* story that alleged some Lumber Liquidators flooring products made in China contained potentially dangerous levels of formaldehyde, the company's stock dropped by roughly two-thirds. Lumber Liquidators' CEO resigned and the company faced an onslaught of litigation (Wilson and Townsend, 2016).

## Sustainability can drive innovation

It is no coincidence that many of society's most innovative products and services (LED lighting, occupancy controls, hybrid and electric vehicles, etc.) also contribute to improved sustainable performance. Likewise, the most innovative production processes (logistics optimization, remanufacturing, green chemistry, etc.) improve the sustainability performance of operations. Countless opportunities exist to incorporate advances in information technology, sensors, nanoscale materials, and biotech into conventional industrial technology to create step-change advances in resource productivity that can expand markets and profit pools. The next generation of engineers and scientists currently studying fields related to science, technology, engineering and mathematics (STEM) in high school and college consider sustainability and innovation to be nearly synonymous. They can't fathom why anyone would bother with an innovation if it didn't improve sustainability performance. What would be the point?

# 8

# Sustainable innovation

> Wind extinguishes a weak fire but it strengthens a healthy blaze.
>
> Tim Lindsey

For many people, sustainability is a philosophy that envisions a more enduring and equitable way of life. The term can have an almost spiritual meaning for some individuals and can evoke considerable emotion and passion. They envision and debate the various aspects and priorities associated with what a more sustainable world would, or should, look like. Their visions of a sustainable world tend to be guided primarily by their personal values and preferences. These perspectives speak to the importance of the concept but are not particularly helpful when it comes to identifying and implementing the changes needed to improve performance. Just because people are passionate about the importance of a concept such as sustainability does not mean that they can, or will, take the substantive actions needed to drive it. Individuals must have a good understanding of the "how-to" aspects of sustainability improvement before they will take action to drive change. In my experience, this is best accomplished by approaching sustainability as an innovation because **innovation frequently provides the catalyst that enables effective change.**

Innovation is the process of introducing and implementing new ideas, methods, or devices. Sustainability can be regarded by most organizations as an innovation in and of itself – an innovative way of doing business. Much like the assembly line, e-commerce, or "lean" management systems, incorporating sustainability principles into the way an organization conducts business can require considerable disruptive change. Instead of focusing strategy predominantly on quarterly economic performance, organizations guided by sustainability principles adjust their focus toward longer-term strategies that balance economic, social, and environmental performance. This change in strategy requires a major shift for most organizations, and accomplishing such a shift requires considerable leadership and commitment to change.

In response to growing demands for more sustainable processes and products, a new class of innovations is emerging. Unlike early industrialization when innovations focused primarily on improving the productivity of people, new innovations produce additional benefits associated with improving health and safety and maximizing the productivity of other resources such as energy, materials, water, and land. Additionally, these more sustainable innovations prevent the generation of wastes, emissions, and pollution. Examples of such innovations include hybrid vehicles, renewable energy, recycling, and remanufacturing systems. Additionally, the "internet of things" is increasingly contributing to improved productivity and efficiency by using sensors and digital technology to effectively connect components in ways that optimize the performance of systems.

FIGURE 8.1 **Stages in the innovation adoption process**

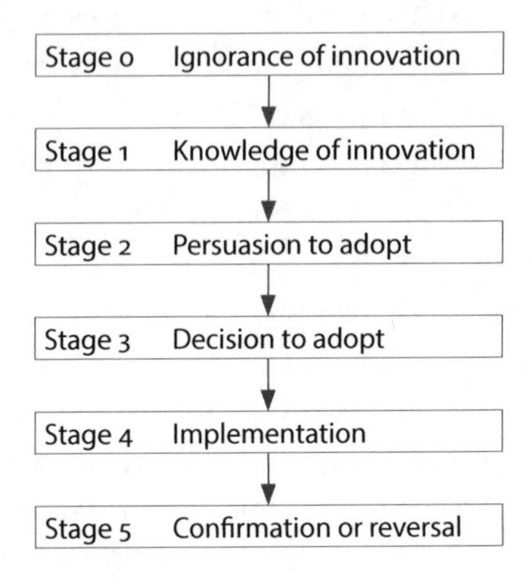


Source: Adapted from Rogers, 2003.

According to Rogers (2003), adopters of innovations generally pass through six stages in the adoption process (see Fig. 8.1). Beginning at Stage 0, a state of ignorance, an individual proceeds to Stage 1 and receives knowledge of an innovation. A period of persuasion (Stage 2) then follows during which the individual recognizes a potential application for the innovation and seeks information to reduce uncertainty about the innovation. Finally, in Stage 3 a decision is made either to adopt or reject the innovation and in Stage 4 the innovation is implemented. This may be followed by a confirmation Stage 5 in which the decision may be affirmed or reversed. In reality, this process may not always be linear. In particular, it is likely that an iterative process between Stages 1 and 2 must occur before a decision is made to adopt the innovation.

## Innovation characteristics that affect adoption rates

In addition to being an innovative way of doing business, sustainability is also a cluster of individual innovations that include ideas, methods, software, devices, machines, materials, chemicals, and energy sources. The adoption rate of each of these innovations is constrained by principles of innovation diffusion. For example, the characteristics of innovations, as perceived by adopters, can affect their rate of adoption. Rogers (2003) describes five characteristics that affect the rate of adoption, as shown in Table 8.1.

**Table 8.1 Innovation characteristics identified by Rogers (2003) and their relevance to sustainable innovation**

| Innovation characteristic | Relevance to sustainable innovation |
|---|---|
| **Relative advantage:** The degree to which an innovation is better than the idea it replaces | Benefits accrue through better safety, efficiency, cost, and revenue performance along with waste reduction, emission reduction, and community benefits |
| **Compatibility:** The degree to which an innovation is consistent with existing values, past experiences, and needs | Connecting innovations with existing culture, values, norms, and priorities can facilitate adoption |
| **Complexity:** The degree to which an innovation is perceived as difficult to understand and use | Innovations that do not require additional skills and expertise tend to be more readily adopted |
| **Observability:** The degree to which the results of an innovation are visible to others | Innovations that can be easily demonstrated, observed, and understood tend to be more readily adopted |
| **Trialability:** The degree to which an innovation can be experimented with on a limited basis | Innovation trials can address uncertainty and risk by providing users with a "hands-on" opportunity to validate performance and address complexity and compatibility issues |

Many advocates of specific innovations assume that if an innovation possesses strong **relative advantages** over the idea it supersedes, diffusion into the market will proceed at a brisk pace. This assumption is usually not valid because, in reality, all five of the characteristics described above are important with respect to adoption rates. A mobile phone is a good example of an innovation that is strong with respect to all five characteristics. However, this was not always the case. I knew a fellow in the mid-1970s who owned a cellular telephone which he kept in a large bag in his vehicle: it was expensive, it was complicated to use, it wasn't very compatible with the conventional phone network, and annoyingly his horn honked whenever he received a call. Consequently, widespread adoption of mobile phones did not occur until the deficient characteristics were improved. Likewise, many sustainability innovations offer strong relative advantage but, if deficiencies associated with the other four characteristics are not addressed, adoption will likely be slow.

## A disruptive approach to conducting business

Taking steps to improve and balance an organization's economic, environmental, and social performance can be disruptive and challenging for many organizations. Strategies and tactics that enable this approach may not be consistent with conventional management priorities and the short-term metrics used to track and assess performance. In fact, guiding an organization toward this approach can seem reckless to an executive who is regularly held accountable by a board of directors for achieving goals and targets that are focused on short-term financial performance. Most executives recognize that sustainability is important to their organization's future and they want to make progress toward improving performance, but they will also continue to be judged on short-term results.

Organizations have a tendency to "cherry-pick" sustainability-focused opportunities that are closely aligned with business as usual. This allows them to demonstrate that they are making progress without significant disruptions to "business as usual." They frequently pursue opportunities associated with improving efficiency and productivity because these approaches achieve short-term cost reductions while also generating social and environmental benefits. The financial benefits of efficiency and productivity measures are easily demonstrated because the resources that are preserved (e.g., energy and materials) have strong monetary value in existing markets. These types of "low hanging fruit" opportunities are a great way for an organization to start down the path toward improved sustainability performance. They demonstrate that the organization can improve its social and environmental responsibility while also improving its bottom line.

As the sustainability journey of an organization proceeds, the opportunities tend to become more challenging because the business case associated with the benefits becomes less straightforward. For example:

- The benefits from improving performance may not have strong monetary value in existing markets (e.g., the value of clean air and water is not accurately quantified in today's markets)
- The benefits from improving performance are often achieved by preventing unwanted events (e.g., accidents and pollution): it can be difficult to quantify the value of events that don't occur

- The benefits from improving performance may accrue over a long-term horizon that exceeds conventional financial metrics
- The benefits from improving performance have considerable monetary value but the benefits may not accrue directly to the organizations that invest (e.g., investing in community or environmental improvements may be a great way to improve the quality of life for employees, and it can increase property values, but it doesn't directly generate revenue for the investing organization)

These challenges can make it difficult to convince leaders that escalating their commitment to sustainability is prudent, especially in the short term. However, progress is possible because, in addition to achieving financial targets, leaders are also charged with developing vision and strategy for ensuring the long-term vitality of organizations. To that end, they regularly consider strategic options and directions for achieving enduring growth and prosperity.

Sustainability can be an extremely complicated subject with countless meanings, aspects, and priorities for the organizations and individuals who pursue it. The values and preferences of diverse stakeholder interests add additional complexity with respect to priorities and methods. It is important to take the mystery out of sustainability and talk about it in terms that are relevant to all aspects of the organization. Many leaders perceive sustainability to be important, but not core to the mission of the organization. They often delegate issues and tasks associated with sustainability to ancillary units, such as government affairs, environmental health and safety, or communications. Organizations that take this approach are unable to capitalize fully on the importance of the topic and the opportunities it can create. Employees in every department of an organization, at all levels, can contribute to improving the long-term viability of their organizations. Sustainability needs to be a meaningful part of everyone's job, similar to safety and quality.

Operating more sustainably can improve an organization's effectiveness but responsibility for sustainability implementation and performance can't be confined to a single department or team. No responsible organization would consider forming a team who has sole responsibility for making the organization profitable. Culture that operates profitably must be woven throughout the fabric of an organization. All departments and employees must understand how to operate profitably and how they can contribute to, and support, the overall profitability of the organization. Likewise, an organization's sustainability strategy needs to engage all levels, specialties, and geographies of an organization. All departments and employees need to understand how to operate more sustainably and how they can contribute to and support the long-term endurance and resourcefulness of the organization.

## Making sustainability compatible with culture

> Look at the ships also, though they are so great and are driven by strong winds, are still directed by a very small rudder wherever the inclination of the pilot desires.
>
> James 3:4

As an innovative business strategy, sustainability tends to present challenges for many organizations with respect to its compatibility and complexity characteristics. Given that many organizations are driven predominantly by short-term financial performance, changing their priorities to give equal emphasis to environmental and social aspects can be problematic. While they may recognize the importance of these aspects, giving them equal emphasis to short-term financial performance is not compatible with the culture of most organizations. Fortunately, many organizations already assign paramount importance to some nonfinancial aspects that can be leveraged to make the case for elevating the status of comprehensive sustainability performance.

Even leaders of the most conservative organizations regularly make decisions for the good of the enterprise that do not demonstrate strong short-term financial benefits. They recognize that certain decisions are in the organization's best interest in the long term even if it is not possible to justify them through the calculation of a short-term return on investment (ROI). Safety and quality are good examples of performance aspects that require long-term focus and commitment. Even though they are difficult to connect with short-term financial performance, most organizations recognize their importance for achieving enduring success. No responsible leader would disregard their organization's safety or quality performance. The same rationale can be used with respect to other dimensions of sustainability. The relationships between sustainability, safety, and quality are described in detail in Chapter 11.

**Case study: Making sustainability compatible with culture at Caterpillar Inc.**

In February of 2012, I was hired by Caterpillar Inc. to serve as their Global Director of Sustainable Development. Caterpillar is the world's leading manufacturer of construction and mining equipment, diesel and natural gas engines, industrial turbines, and diesel-electric locomotives (Caterpillar Inc., 2017). Caterpillar leaders made sustainability part of the enterprise strategy in 2005, identified it as a growth area and launched their first Sustainability Report. When Doug Oberhelman became Caterpillar Chairman and CEO in 2010, he clearly had a passion for sustainability and was focused on integrating it into the business. Many Caterpillar executives knew intuitively that the sustainability movement presented strong opportunities for the company because of the products it makes and the markets it serves, but, they weren't in agreement about how to capitalize on this opportunity. I knew some of their leaders through projects I had directed for them in the past and they recruited me to take on this challenging position.

From the beginning, I approached sustainability at Caterpillar as an innovation – an innovative way of doing business. My team of 2–3 personnel and I began by researching key aspects of the culture that we thought could connect well with sustainability. We studied the business model, the quality system, manufacturing operations, logistics, the human resources system, product development, the supply chain, customer relations, communications, strategy, and the corporate values. We concluded that all these aspects offered considerable potential for improving the sustainability performance of the enterprise.

During my first day on the job, I noticed that the company placed considerable emphasis on its four core values: integrity, commitment, excellence, and teamwork. These values formed the basis of Caterpillar's Worldwide Code of Conduct and were considered in all major decisions. Posters that describe and emphasize the values were displayed throughout the company's facilities. As I was making the rounds and meeting with various Caterpillar executives during my first days there, one of them asked me what I thought we needed to do to "raise our game." I responded, "If we are really serious about sustainability, we should consider making it a core value." Such a change could potentially be transformational. It would mean that sustainability would no longer be just important things that we do – it would become part of who we are.

To be effective in this change, I would need to enlist the influence of key leaders who could help champion the new emphasis on sustainability to their respective business units and help drive the transition. I was extremely fortunate when Caterpillar Chairman and CEO Doug Oberhelman assigned a special leadership development team to investigate sustainability opportunities for the enterprise and develop recommendations for elevating Caterpillar's sustainability commitment and performance. The leadership development work was part of a program with Stanford University and the team was comprised of eight very accomplished leaders representing diverse aspects of the organization, including marketing, investor relations, logistics, manufacturing, and product development. I was assigned the role of serving as their subject matter expert.

Over the course of nine months, we met multiple times and spent many hours together. After several months of benchmarking other companies, information gathering, and study, the team had developed a deep understanding of the topic. They recognized Caterpillar's unique opportunity to lead in solving some of the world's most pressing problems, ranging from access to energy, transportation, water, and sanitation, to restoring degraded lands and sequestering greenhouse gases. After some thoughtful discussion, the team reached consensus. If Caterpillar wanted to be a global sustainability leader, it made sense to recognize sustainability as a core value.

The recommendation to make sustainability a core value was presented to top management in the form of a carefully crafted video that made the business case. The video also connected at an emotional level and emphasized the world's need for Caterpillar leadership. Top management agreed, and the change was presented to the board of directors who supported and approved the measure.

An implementation team was then formed to carefully craft the language to support the value while maintaining compatibility with the other values. We developed compatible messaging regarding the meaning and importance of sustainability to the enterprise. In addition, an internal steering committee was formed with management representatives from strategy, quality, product development, human resources, communications, and purchasing to develop a coordinated approach for embedding sustainability from the top down throughout all aspects of the enterprise.

During this time of rapid transition, the company doubled its commitment to aggressive energy efficiency and greenhouse gas reduction goals (50% reduction by the year 2010) and implemented sustainability principles in ways that began to drive innovation and improved competitiveness throughout the corporate value chain. Discussions also began about how to incorporate sustainability into corporate roles (research, product development, strategy, and risk) and operational functions (supply chain, manufacturing, quality, logistics, and customer processes).

The corporation officially recognized sustainability as the fifth core value for the enterprise in 2014 and incorporated it into the employee Code of Conduct in 2015. We developed a training module to help employees understand how to embed sustainability principles into their daily work. A "Sustainability Advocates" program was also established with open enrollment for all employees to facilitate sustainability innovation, education, and implementation from the bottom up.

Many transformational changes were implemented over the next two years. As a result, in 2015, 18% of Caterpillar's reported sales and revenues were from products, services, and solutions that demonstrated an improved sustainability benefit over existing offerings. More recent additions include microgrid technology comprised of solar panels and battery storage integrated with Cat® generators to provide power to remote parts of the world, and hybrid excavators that utilize the machine's hydraulic system to store and deploy previously wasted energy. The company also worked with NGOs, engineering companies, financial institutions, and peer companies to develop an initiative focused on restoring natural infrastructure (Lindsey, 2015).

By the time I left Caterpillar at the beginning of 2016, the company had achieved elite recognition for its sustainability achievements, including:

- Top 4% ranking for its sector on the Dow Jones Sustainability Index (after being removed from the index the year after I arrived)
- Runner-up status for circular economy achievements ("The Circulars") at the 2016 World Economic Forum
- The Keep America Beautiful 2015 "Vision for America" award

People often ask me how we achieved such transformational change in this relatively short period of time. I attribute our success to three aspects:

1. We engaged champions with the courage and influence to drive change
2. We found ways to make sustainability improvement compatible with the existing culture
3. We used business drivers to catalyze change

## Dealing with complexity

Some sustainability issues tend to be extremely complex due to their technical, social, and political nature. Diverse stakeholders, representing a wide variety of vested interests, emphasize aspects of the issues that are most favorable to their agendas. Consequently, consensus regarding the best strategy for addressing the issues can be difficult to achieve.

Nowhere is complexity more evident than the climate change issue. A variety of solutions are available that can potentially mitigate climate change. However, considerable complexity exists with respect to both the climate change issue itself, and the potential solutions available to address it. Many activists and policy-makers frequently explain that the climate change issue is a consequence of greenhouse gas accumulation, resulting from the combustion of fossil fuels. However, other interests are quick to point out that climate is also affected by changes within the sun, in the Earth's orbit, and in the reflectivity of the Earth's surface, and by particles suspended in the atmosphere (EPA, 2017d).

Even when stakeholders agree that greenhouse gas accumulation is a primary concern, the relative sources of greenhouse gas are often disputed. Many stakeholders focus almost exclusively on emissions from fossil fuels while others point out that additional sources of greenhouse gas emissions are equally, if not more, important. Billions of tons of greenhouse gas are produced annually from soil tillage, livestock, deforestation, wetland destruction, food waste, and waste disposal.

With so much confusion over the sources of climate change and greenhouse gases, it is no wonder that organizations have so much difficulty agreeing on which innovative solutions will most effectively address the problem(s). For instance, countless advocates for renewable energy suggest that switching from fossil fuels to wind and solar will fix the climate problem even though these technologies will not address the nonfossil fuel sources of greenhouse gases. Plus, widespread wind and solar implementation can be challenging because of intermittency issues associated with daylight and weather. Additionally, their total costs can be confusing because of the complex system of subsidies that vary from region to region, and emerging options for storing energy to address the intermittency issues. To further complicate the matter, other options for power generation exist that can reduce greenhouse gas emissions with fewer intermittency issues, including nuclear power, combined heat and power, and additional sources of renewable energy (geothermal, biomass, and hydropower). Each of these alternatives contributes further complexity to the issue through its specific advantages and challenges. When political agendas and aggressive lobbying measures are added to the mix, overwhelming levels of uncertainty accelerate the constraint headwinds and progress lags.

A great deal of investment and progress has been stifled by constraint headwinds driven by uncertainty associated with the future of the climate change issue. Some governments are attempting to regulate $CO_2$ emissions while others push for various types of carbon tax. Still others promote cap and trade systems for emissions. The strategies for effectively dealing with these contrasting measures differ greatly. Consequently, many businesses wait in limbo until they receive more definitive direction from government.

The high levels of complexity and uncertainty associated with the climate change issue are typical of many sustainability aspects. The processes and systems involved are complicated, and new innovations can affect diverse stakeholders positively and/or negatively. Addressing the specific information needs of the various vested stakeholders is critical for ensuring the successful adoption of many innovations.

## Knowledge requirements for more sustainable innovation

**When curiosity and imagination collide, they spark creativity. Creativity sparks ideas and ideas lead to innovation.** Innovation often involves a significant amount of uncertainty and risk because innovations don't always perform as planned and they are frequently disruptive to an organization's culture and operations. To reduce uncertainty and risk to acceptable levels, decision-makers often have to gather considerable information and knowledge regarding the innovations under consideration. Rogers (2003) suggests that an adopter must possess three types of knowledge about an innovation before they decide to adopt it. These three types of knowledge are described briefly below, along with examples of information sources that are commonly used to address the knowledge gap.

1. **Awareness knowledge:** Information that an innovation exists and may warrant further investigation
   *Information sources:* Websites, posters, trade journals, workshops, and seminars
2. **Principles knowledge:** The underlying technical aspects that determine how an innovation functions
   *Information sources:* Experts, handbooks, manuals, textbooks, training, and academic courses
3. **"How-to" knowledge:** Information needed to implement and use an innovation productively in a specific application
   *Information sources:* Pilot trials, demonstrations, peers, vendors, and consultants

Many advocates of more sustainable innovations focus on creating awareness associated with the many advantages that can result from implementation. The rationale behind this approach is based on two assumptions:

1. If potential adopters can be made aware of the innovation's many advantages, they will proceed with adoption

2. Awareness information will be adequate for adopters to make an informed decision and proceed with implementing the innovation

For most innovations, neither assumption is valid. Awareness information simply is not adequate to address all the uncertainty associated with most innovations. At best, awareness information will heighten a potential adopter's interest to a point where they will consider pursuing additional information.

Understanding the technical principles behind how an innovation functions is crucial before a potential adopter will seriously investigate the feasibility of an innovation. Fortunately, most organizations have access to technical experts either internally or through external sources, who can assess the technical and economic merits. The biggest challenge from an information perspective usually comes down to the "how-to" knowledge required for implementation. Even though decision-makers may possess a strong technical understanding of an innovation's function and benefits, they are often reluctant to proceed with implementation because they are uncertain of how it will function in their specific application.

## Keys for successful implementation

Successful implementation of many innovations requires involvement from champions who exhibit strong leadership, accountability, authority, and access to needed resources. Ensuring that these champions understand how to integrate sustainability principles into their operations in ways that are compatible with existing culture is paramount to success. Site-specific "how-to" information is often needed to address uncertainty regarding "how to" successfully implement an innovation. How-to information is usually more difficult to acquire than awareness and principles information because it cannot be obtained from mass media sources or technical references. Communications with peer organizations (including competitors) who have successfully implemented the innovation can be very helpful for reducing uncertainty.

Pilot trials are often an effective way to address the uncertainty that accompanies many innovations because they give champions the how-to information that they need to resolve compatibility and complexity issues. By testing the innovation on a limited basis before making a permanent decision to implement, performance data can be collected and used to develop design parameters, and estimate costs and benefits for full-scale implementation. Lindsey (2000) conducted a study of 53 manufacturing companies interested in using ultrafiltration membrane technology to recycle industrial fluids. None of the 47 participants who received awareness information only (fact sheets and case studies) regarding the technology proceeded with implementation despite the promise of economic and environmental benefits. In contrast, four out of six companies that conducted pilot trials of the technology implemented it on a permanent basis.

Market dynamics are often shifted by innovations that affect changes in revenue, costs, incentives, and obstacles. This can be particularly true for innovations that improve sustainability performance. Disruptive changes often produce winners and losers (e.g., the effects of climate change regulation on the coal industry). The key to improving sustainability performance is to pick innovations that create winning outcomes for the economy, communities, and the environment, and losing outcomes for wastefulness, deficiency, dysfunction, and negligence. Directional guidance based on sound principles is needed to ensure that leaders consistently make sound decisions regarding their efforts to improve sustainability.

# 9

# Establishing a sustainability compass

> Sometimes in the wind of change, we find one true direction.
>
> Unknown

The breadth and complexity of sustainability aspects and constraints can be overwhelming and can make it difficult for leaders to develop strategy, priorities, and initiatives. To help make this task workable, I encourage organizations to think of sustainability as a journey. As with any journey, a roadmap could be helpful to facilitate establishing routes, vehicles, schedules, and so on. Unfortunately, when it comes to sustainability, no such map exists that can be applied broadly to organizations. Very few organizations will become fully sustainable during the tenure of current leadership because fully sustainable options for processes, products, materials, energy, and so on are not currently available. Future prospects for more sustainable options are promising but specific choices are unclear because technology shifts are increasingly disruptive to existing social, economic, and environmental systems.

Thinking of sustainability in terms of *nouns* such as "society," "environment," and "economy" (or "people," "planet," and "profit," if you prefer) can provide the rationale behind "why" we strive to operate more sustainably. We want to be effective stewards of these assets, and focusing on their protection and improvement can help organizations identify their biggest impacts and opportunities. However, these nouns provide little insight regarding "how to" go about actually improving performance. Action is required to effect change and this can only be accomplished through work. We use *verbs* to describe work and, in the case of sustainability, verbs such as "prevent," "improve," "optimize," and "restore" can go a long way toward describing the work that is needed for improvement.

## Doing things right versus doing the right things

The development of detailed, long-term, prescriptive plans for becoming fully sustainable is often an exercise in futility. These plans can help establish a common understanding of how to "do things right" but they are usually not effective tactics for ensuring that organizations are "doing the right things." Directional plans based on sound principles of sustainability are often more productive because they allow flexibility when adjusting to technology shifts and changing economic, social, and environmental conditions. Undertaking this journey on a directional path toward becoming "more sustainable" is usually more productive than following a prescriptive plan. The more directional approach is also more conducive to widespread participation because it facilitates engagement with individuals in all aspects, geographies, and roles of the organization. Once everyone is moving in the same direction and following the same principles, then more tactical and prescriptive plans can be developed and implemented to address specific issues and opportunities.

Constraint headwinds can best be defeated when they are approached as an undesirable *stimulus* comprised of problems that require a proactive *response*. The problems need to be analyzed to determine their sources before effective responses can be crafted to address the root causes. To help leaders address constraint headwinds with methods that improve the sustainability of their enterprise, I encourage them to approach sustainability (this method can actually be applied to any complicated issue) using a technique suggested by Covey (2014). Covey suggests that following time-tested principles can ensure that you are always headed in the right direction. If you were completely lost in a jungle and in need of some guidance, what tools would be most helpful to ensure you choose the right direction? A roadmap wouldn't be particularly helpful because you don't know where you are or where you are headed. A more effective tool in this situation would be a compass that would help guide you in the right direction.

As described in Chapter 4, four basic types of root cause can usually be found at the sources of constraint headwinds:

1. Wasteful practices
2. Deficient processes and products
3. Dysfunctional interactions
4. Negligent decisions

Each of these root causes can be defeated by incorporating sound sustainability principles into the management, design, production, and operation of processes, products, and systems.

## Guiding principles for improving sustainability

> If you surrender to the wind, you can ride it.
>
> Toni Morrison

Principles are fundamental truths or propositions that serve as the foundation for a system of belief, behavior, or chain of reasoning, and they guide actions to establish acceptable norms. In his international best-selling book *The Seven Habits of Highly Effective People*, Stephen Covey (2014) argues that "there are principles that govern human effectiveness – natural laws in the human dimension that are just as real, just as unchanging and arguably *there* as laws such as gravity are in the physical dimension." Covey focused on principles of personal growth and change, including fairness, integrity, honesty, dignity, and service – to name but a few. It is difficult to argue against the reasoning behind these principles given their universal acceptance from moral and religious perspectives. Hence, these principles have stood the test of time, and can be regarded as "sustainable" in their own right.

One could argue that living and working sustainably is another example of a principle that should "govern human effectiveness." By following guiding principles for improving sustainability performance, we can establish a source of directional guidance to inform strategy and tactics. Likewise, we can use the principles as a compass to guide the way we develop, design, and manage an organization's processes, products, and systems.

Four guiding principles are described below that can be followed to defeat the root causes of constraint headwinds. In some cases, following these principles may change an organization's course so dramatically that the problematic headwinds actually become a source of beneficial tailwinds. To use the sustainability principles to guide strategy effectively requires broadening our perspectives with respect to how we consider and manage waste, quality, systems, and degraded assets. The principles, and how they can be applied, are briefly:

1. **Prevent waste** hazards and inefficiency that can lead to accidents, injuries, illnesses, unrealized potential, pollution, degradation, and compliance issues
2. **Improve quality** of processes, products, communities, and the environment by correcting deficiencies that can result in waste and hazards
3. **Optimize systems** to ensure that each component is appropriate, functions correctly, and is properly integrated with other aspects
4. **Restore value** to community, environmental, and individual assets damaged from previous negligence

More detail regarding each of these principles and how they can be applied effectively is provided in Chapters 10–13. They are described in the language of business and are directionally compatible with the long-term plans and strategy of virtually all forward-thinking organizations. Each of the four principles is structured to target and defeat a specific type of root cause that contributes to constraint headwinds. Figure 9.1 shows how the guiding principles can be applied to defeat the individual root causes.

Regardless of their status or role, all employees can contribute to sustainability performance by incorporating these four guiding principles into their daily work. To use these principles effectively, employees should consider the long-term risks, challenges, and opportunities that confront the organization with respect to waste, quality, and systems, and how they affect their organization, communities, and the environment. The principles can be applied to

FIGURE 9.1 **How sustainability principles defeat the root causes of constraint headwinds**

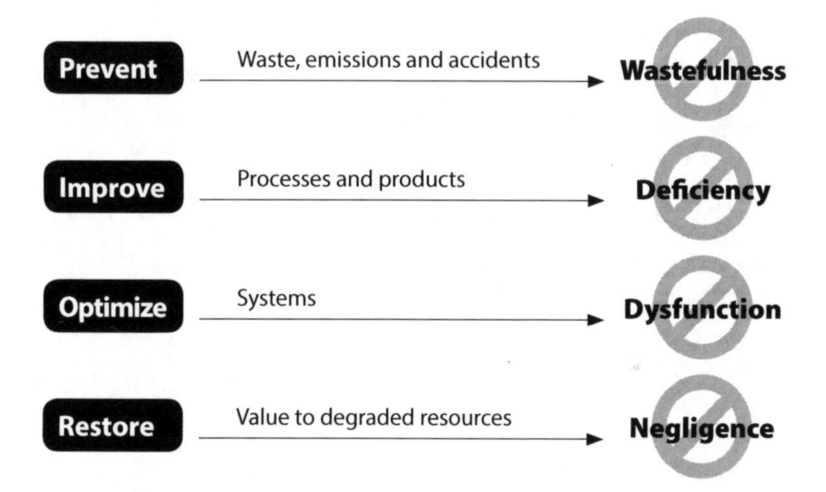


virtually any aspect of work to improve sustainability performance. Examples of aspects where these principles can be applied successfully include strategy, design, methods, talent, teams, machinery, materials, energy, communications, relationships, logistics, innovation, and technology.

When constraint headwinds are approached proactively, using methods guided by these principles, effective measures for improving system sustainability can be identified and implemented more easily. Opportunities for innovation frequently emerge that can actually provide sources of cost reduction, market expansion, economic growth, and competitive advantage while simultaneously improving community health and protecting the environment.

The next four chapters describe how to apply the guiding principles in more detail, and provide examples of how various organizations have used them to improve their sustainability performance and overall effectiveness.

10

# Sustainability principle #1: Prevent waste

> Adversity is like a strong wind. It tears away from us all but the things that cannot be torn so that we can see ourselves as we really are.
>
> Arthur Golden

Throughout my career, I have heard businesses repeatedly complain about the high cost of managing their wastes. The argument usually goes something like this: "We can't compete with [insert the name of a country from the developing world] because they are allowed to dump their waste in a river, whereas here in the U.S.A. we have to follow stringent regulations. It just isn't fair." I have come to realize that this argument is not valid or appropriate for modern enterprise. It is equivalent to saying, "We can't compete with the developing world because they are allowed to be more wasteful." The logic behind this argument just doesn't wash. Certainly, managing wastes associated with by-products from production can be challenging. However, waste is usually a symptom of deeper problems that can be addressed more effectively with measures that eliminate waste at its source. Such measures are much more effective than dumping waste in a river or burying it in a landfill.

Benjamin Franklin is credited with coining the phrase "an ounce of prevention is worth a pound of cure" (Franklin, 1999). The enduring wisdom of this viewpoint can be applied across all types and sizes of issues. My more than 35 years of experience with helping organizations of all types and sizes become more sustainable has led me to an incontrovertible truth regarding the relationship between waste and sustainability – **that organizations become more sustainable as they become less wasteful.** This applies broadly to diverse organizations, including individual households, businesses, communities, regions, nations, and planets.

Figure 10.1 shows the relationship between sustainability and wastefulness. Waste is created through either the degradation or consumption of all types of resources. Degradation occurs when

FIGURE 10.1 **The relationship between sustainability and wastefulness**

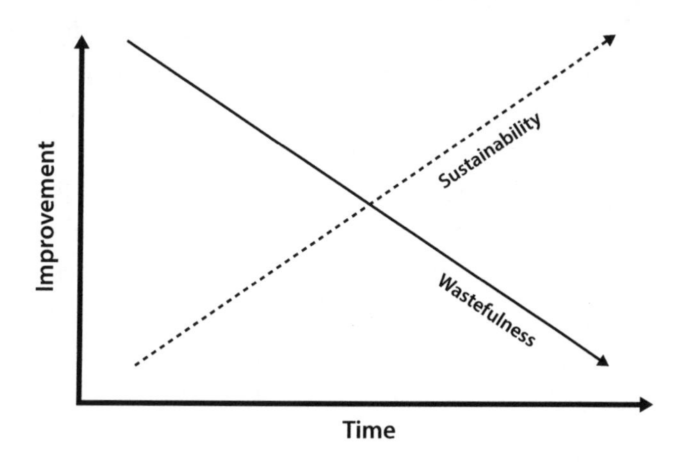

resources are damaged or contaminated by accidents or pollution. Consumption can be wasteful when raw materials, such as energy, water, metals, chemicals, paint, plastic, and wood, are converted into products and by-products that permanently leave the value chain. The resulting unusable waste and emissions can be harmful and can create additional waste by degrading other resources. Human resources can be wasted when individuals do not have opportunities to achieve their full potential through education and gainful employment, or when they experience health problems from unsafe practices and conditions.

Wastefulness can occur in multiple forms and is frequently an outcome of deficient processes and products, and dysfunctional systems. It can deplete the availability of materials and energy and reduce the value and capabilities of agricultural, industrial, community, and natural resources. Finally, wastefulness can reduce the ability of current and future generations to achieve a good quality of life. Preventing all types of waste provides multiple benefits for the economic, social, and environmental dimensions of sustainability. Prevention of waste can be accomplished through various measures that address inefficiency, hazards, accidents, attrition, defects, by-products, and excessive resource consumption. Table 10.1 provides examples of wastefulness impacts that can be addressed by proactively taking measures to prevent waste. Preventing wastefulness through safer and more productive use of all resources is critical for improving sustainability performance. The benefits that can be achieved through preventing all types of waste are endless, and employees throughout an organization can contribute through a variety of continuous improvements that are common in many organizations.

**Case study: 3M "Pollution Prevention Pays" program**

3M is widely recognized as the first U.S. corporation to aggressively implement preventive strategies to reduce waste and toxic releases. In 1975, 3M initiated its "Pollution Prevention Pays" program in response to the environmental legislation and regulations of the early 1970s, which specified increasingly sophisticated and expensive treatment technologies as control measures. 3M placed emphasis on preventing the generation of pollution at the source through a combination of product reformulation, process modification, equipment redesign, and waste recycling and/or reuse. Since the program's inception, 3M has undertaken over 8,100 pollution prevention projects that reduced its waste generation by over 2 billion tons and saved the company over US$1.9 billion (3M, 2017).

By increasing the safety, efficiency, and productivity of processes and products, costs can be lowered from reductions in the use of materials, energy, water, and land. Providing a safe work environment and the tools and training that employees need to work safely are crucial for preventing waste. Nothing is potentially more wasteful than an accident – particularly if someone is injured – or if damages occur to community assets, company assets, or the environment. This applies to both employees and to customers that use an organization's products and services. Keeping employees engaged is another key to preventing waste. Employees should have opportunities for professional growth and career development to reach their potential and make long-term valued contributions.

Although prevention-based strategies have been proven time and time again to be the most effective approach to dealing with all types of waste and hazards, many organizations still respond to wasteful constraint issues in a reactive, linear fashion. They frequently accept the wasteful issues they face as a necessary evil and do not attempt to uncover underlying problems. In most cases, reactive responses to addressing wastefulness do not produce the most effective solutions for addressing constraints because they do not address the root cause of the problem. More effective approaches involve analyzing the processes and products that create the wasteful issue, including an assessment of the various inputs and outputs associated with individual aspects and steps. The relative quantities and costs associated with the inputs and outputs can often be calculated to facilitate full accounting of the wasteful practice.

Once the processes and products that contribute to wastefulness are fully understood, an accurate assessment of the problem's root cause(s) can be determined and appropriate solutions can be developed. In many cases, problems can be corrected through relatively simple, incremental changes. However, more complicated problems may require significant process modifications and adoption of innovations that are not common to the organization.

Figure 10.2 demonstrates the basic flow of activities that are commonly undertaken to identify root causes of problems and develop effective solutions. By using sound investigative techniques to follow the process described in this diagram, practitioners can help ensure that the correct problems are addressed using the most effective means available.

Trials of innovations may be needed to assess their potential impacts before committing to full-scale implementation. In some cases, technology may be available to address the problems but it has not yet become commercially available and some testing will be needed to assess the technology and determine if modifications are needed to match the application. Finally, some problems simply have no viable solutions at present and will require research to address them.

No prescriptive checklist or smartphone app is available that can lead individuals to the root causes of problems or to the best

**Figure 10.2 Activity flow for identifying and addressing the root causes of wastefulness**

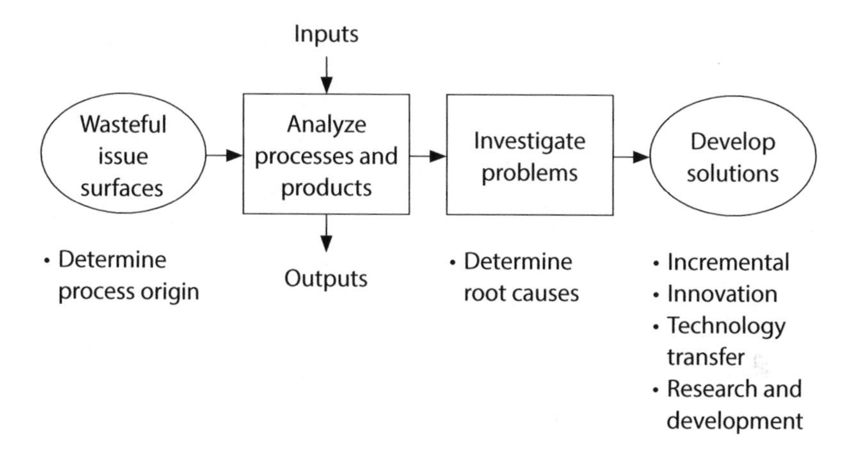


solutions for addressing them. These issues have to be approached from the perspective of a detective who follows the evidence where it takes them. Multiple training programs are available to help individuals learn effective techniques for performing root-cause analysis.

Table 10.2 describes some common examples of wastefulness problems that regularly occur in business operations and examples of responses that are commonly used to address the problems. Both reactive and proactive responses are shown to illustrate the difference between reacting passively to problems versus addressing them proactively to address the root cause.

Each of the proactive responses is focused upstream at addressing the source of wastefulness, and the outcomes are considerably more effective. The measures associated with the proactive responses will all lead to gains in efficiency, productivity, and safety, whereas the reactive approaches described in the middle column would lead to additional cost and complexity.

**TABLE 10.2 Examples of reactive and proactive responses to wastefulness issues**

| Perceived issues | Reactive responses | Proactive responses |
|---|---|---|
| 1. Volume and cost of waste disposal is high | Find a lower cost landfill | Modify processes to prevent waste creation |
| 2. Treated waste-water exceeds permit limits | Find a more effective waste-water system | Modify processes to keep contaminants out of water |
| 3. Work process is dangerous | Outsource process work to a vendor | Modify process to eliminate hazards |
| 4. Community objects to new site development | Relocate to a more cooperative region | Investigate potential use of brownfield sites |
| 5. Workforce lacks technical skills | Bring in skilled workers from other regions | Work with community colleges to provide appropriate training |

### Case study: Identifying the source of oily waste-water at railroad maintenance facilities

This case study describes how some specific wasteful practices were addressed at locomotive maintenance facilities. It was selected because the operations are relatively straightforward and easy to understand – similar to an automotive maintenance facility where you might have your vehicle serviced. The types of practice described in this case study are not unique to railroads. I have encountered similar wasteful practices in hundreds of operations across multiple sectors. In my experience, the railroad industry is actually better than most when it comes to implementing proactive waste prevention measures.

In the mid-1990s, I was contacted by a large railroad enterprise about a waste-water issue they were experiencing at a facility they used to service locomotives in downstate Illinois. We had worked with them quite a bit and they were comfortable sharing details associated with their issues. The local community's sanitary district would no longer accept the discharge from the railroad facility's waste-water effluent because it contained oil in concentrations that exceeded the limits that the community could accept and treat. The railroad was forced to pay a vendor considerable sums to collect the waste-water and haul it off-site to an operation where it was treated using technology that was more effective than the railroad's own waste-water treatment capabilities.

The railroad had struggled with this issue for several years, and in two previous years its environmental engineer had proposed a US$2.2 million capital project to upgrade the facility's waste-water treatment system. Unfortunately, the project competed for capital with other needs and had been rejected each year it was submitted. The project was approved on the third attempt and the environmental engineer was optimistic that he would now be able to find a more effective solution to address the oily waste-water problem. He initially called me to discuss technology options for removing oil from the oily waste-water. His request puzzled me initially because, of course, oil and water don't normally mix very well and can usually be separated with relatively simple oil/water separation technology. He explained that they already had such a separator in place but it would not handle the demands they had placed on it. Still puzzled by the scenario he had described, I suggested that I visit the facility and study the source of the problem before making any recommendations on how to solve it. He agreed and I convinced him to set up a meeting while I was there with the plant manager, maintenance manager, purchasing director, and others that might have insight into the processes and materials that impacted their waste-water.

On arriving at the facility, they gave me a tour of the operations and I marveled at the scale of the shop bays and equipment they used to service and repair locomotives. They also showed me their waste-water treatment system and the milky waste-water that it generated. We adjourned to a meeting room on-site where I asked them to walk me through the various process steps they used to service a locomotive. I recorded their descriptions on a flip chart at the front of the room by drawing a basic process flow diagram that included all the inputs and outputs for each step. Figure 10.3 is a general description of the first two steps of the locomotive servicing process.

Other than the extraordinarily large scale, the process seemed to be fairly straightforward – much like an automotive service facility. As shown in the figure, their standard operating procedure (SOP) directed technicians working on the locomotive to place it in a staging area where they would wash the unit with hot soapy water and empty the trash. Technicians then transferred the clean locomotive to a service area where they drained and replaced various fluids and replaced excessively worn parts.

**Figure 10.3 Locomotive shop process flow**

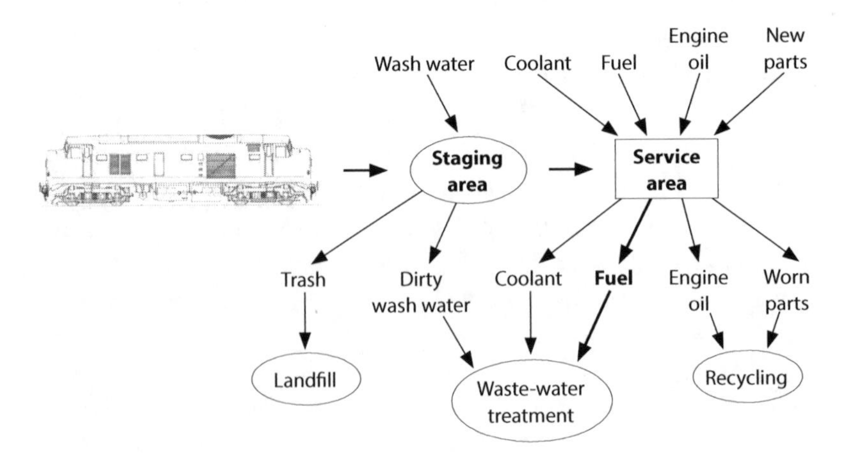

In relatively short order, the developing process flow diagram revealed the source of the oily waste-water problem. The diesel fuel and coolant were drained to the waste-water treatment plant where they mixed with the hot soapy water from the staging area. The plant manager was incredulous when the process details unfolded on the flip chart. "You've got to be ____ting me," he protested, questioning if the process was really performed that way. He pointed out that locomotives can frequently hold hundreds of gallons of diesel fuel which was a major expense for the operation. The maintenance manager not only confirmed that this was indeed the SOP, he explained that it has been done that way for many years while servicing hundreds of locomotives. The facility was designed and built decades earlier when fuel was cheap and the designers did not include the plumbing and tanks needed to enable fuel recovery.

With this newly gained knowledge, an alternative path for addressing the problem became obvious. Instead of investing in expensive waste-water treatment technology, they could simply install a couple of tanks in the shop and drain the fuel into them. The fuel would be transferred back into the locomotives when the service cycle was completed. Since the fuel would no longer be mixed with the hot soapy water, the waste-water would be much cleaner and not require treatment beyond their existing oil/water separator.

By implementing process improvements that addressed the root cause of the problem, this operation was able to prevent waste instead of treating it. In theory, they should have saved big money on fuel while successfully eliminating the need to invest US$2.2 million in an upgraded waste-water treatment system. In practice, it didn't play out quite that way. Of course, they proceeded with implementing the process changes needed to recover the fuel, and this modification did indeed save them a great deal of money. However, they also went ahead with the

US$2.2 million investment in the waste-water treatment system. Why on earth would they invest in equipment that they no longer needed? Well ... they had been pushing for the upgrade to their waste-water treatment system for three years and had aggressively presented it as a high priority. They were finally able to justify the upgrade as a measure to achieve compliance and reduce risk. None of the individuals involved wanted to be the one who explained to top management that they would no longer need the expensive system because they decided to stop dumping diesel fuel into their waste-water.

The story doesn't end there. Given the successful outcome of the Illinois project, the company's management recommended that we investigate a similar issue at a facility in Iowa. Sure enough, the Iowa facility was draining diesel fuel to its waste-water as well. Several months later, we were called to a similar facility in Nebraska that, you guessed it, was using the same wasteful process.

Hang in there with me – the story continues. While participating in a trade association meeting with other railroad companies, the company we worked with on the Illinois, Iowa, and Nebraska facilities recommended our services to a competing railroad. The competing railroad was experiencing oily waste-water problems at maintenance facilities in Texas, Arkansas, and Idaho. We visited all three facilities and, you guessed it, all of them dumped diesel fuel down the drain during their maintenance processes.

This story demonstrates how wasteful practices can become institutionalized in organizations and be repeated hundreds of times over many years. It also shows what can happen when organizations try to address symptoms instead of root causes. One question that frequently comes up regarding this story is "Why didn't the people at Facility X simply tell the people at Facility Y about the diesel fuel issue?" The answer to this question is actually quite simple. Nobody wanted to admit that the source of the oily waste-water problem was the wasted diesel fuel because they couldn't be sure that other facilities followed the same procedures. As far as they knew, the sources of oily waste-water at the other facilities could have been more sensible. By referring the additional facilities to me, they were able to help their colleagues without admitting to their own wasteful practices.

This story seems unbelievable to many people and, on hearing it, they often assume that this type of problem is a freak incident and would never happen at most operations. Sometimes they suggest that railroads must be laggards with regard to environmental protection. This is simply not the case: I have worked with multiple railroads over the years and they have implemented some of the most proactive and comprehensive programs I have ever observed with respect to safety and environmental protection. I have encountered these types of wasteful practice in hundreds of operations across dozens of sectors. If I had chosen an example in another sector, such as metal finishing or chemical processing, the processes would be much more difficult to explain to those who are not familiar with them. However, the examples would be every bit as wasteful.

A great deal of waste is generated simply through administrative and management choices. For example, many organizations have established policies to replace computer hardware with regular frequency. Many companies choose to replace personal computers every three years in an effort to keep everyone's information technology capabilities current. The three-year duration was established 10–15 years earlier, when advances in software technology outpaced advances in hardware technology. Software breakthroughs were common and could only be used with recently updated hardware. In most cases, there wasn't anything wrong with the hardware – it simply lacked the capacity required to use the latest software. Vast amounts of electronic waste have been generated due to this effort to stay current with rapidly changing technology. In recent years, breakthroughs in hardware technology have allowed computers to keep pace with software development such that computers no longer need to be replaced as often. However, many organizations have kept policies intact to replace computers at the same regular intervals they established over a decade earlier when conditions were much different. Such policies institutionalize waste generation by ensuring that perfectly good hardware is needlessly replaced.

## Assessing the full cost of wastefulness

Waste prevention can frequently be accomplished with minimal cost through incremental changes in behaviors and practices. However, in some cases, significant investment in equipment and/or facilities may be required to address the root cause of wastefulness. In order to justify such investments, it is extremely important that the full cost of wastefulness be considered. For instance, costs for managing common industrial wastes can include storage, shipping, treatment, incineration, disposal, record-keeping, and accident mitigation, and they can add up quickly. However, these costs are frequently only the tip of the iceberg. The monetary costs of wasted raw materials and energy frequently dwarf costs associated with managing wastes after they are created. Additionally, it is important to recognize that relatively small magnitudes of hazards, waste, and pollution can degrade much larger community and environmental resources.

Considering the full cost of wastefulness is crucial for prioritizing problems and justifying measures for prevention. Taking measures to prevent waste can result in considerable cost savings when the resources being conserved (e.g., energy and materials) have high market value. However, some resources and services provided by the environment have low monetary value in existing markets (e.g., water and air), even though these resources are essential to the well-being of businesses, people, and the environment.

In addition to supplying public goods, the environment also provides a variety of "ecosystem services" that are not accurately valued in markets. Examples of ecosystem services include:

- Production of food, fiber and water
- Regulation of the climate and disease
- Supporting nutrient cycles and crop pollination
- A variety of cultural, spiritual, and recreational benefits.

An analysis by de Groot *et al.* (2012) screened over 300 case studies on the monetary value of ecosystem services. The average value of ten main ecosystem types was calculated. The total value ranged between US$490 per hectare annually for open ocean, and US$350,000 per hectare per year for coral reefs.

Markets fail to capture the true value of these resources and services because they are perceived to be "public goods" that are equally available to everyone. This factor can be a stumbling block when trying to advocate for measures that can produce considerable social and environmental benefits because the market-based ROI may not be competitive with more conventional project opportunities. Ensuring that resources with low monetary value are not wasted can be one of the most challenging aspects of sustainability leadership.

Several years ago, I was giving a presentation on sustainability principles to a group of corporate accountants. Initially, they didn't show much interest in the topic because they didn't believe it was particularly relevant to their work. Their interest grew a bit when I described the strong business case that the company had realized in recent projects, which had achieved considerable sustainability benefits through improved efficiency with energy and materials. However, their skepticism returned when I began explaining that preserving resources with low monetary value is more challenging. To make the point that these resources are relevant as well, I asked them how much the corporation had paid for air in the previous year. They responded with blank stares so I posed the question again, and once again I received no response. I then posed the question, "Does air have any value?" to which one brave soul responded, "Apparently not, since we don't pay anything for it." I then suggested that perhaps he should try going five minutes without some air, then he would gain an appreciation for just how valuable it is. The accountants acknowledged the point but remained resistant to the idea of modifying their procedures to account for resources that do not have monetary value in markets.

It can be difficult to convince individuals working in the accounting community that they have a critical role in driving sustainability performance. Their previous experiences with measures that improve environmental and/or social performance often leads them to believe that these efforts require significant monetary investment with little return. They can cite instances where this has been the case but their examples may or may not be relevant depending on the issues at hand and the approaches (reactive or proactive) used to address them. Cost-effective measures that utilize prevention strategies can usually be justified when the full costs of the activities, materials, and energy associated with problematic wastes are accounted for.

Although some resources have low monetary value when considered in isolation, the full cost of actually using the resources can be much higher than the cost of simply purchasing them. A full accounting of costs associated with processing, maintenance, additives, storage, waste disposal, compliance, hazards, risks, and future liabilities can provide a more accurate estimate of the full costs associated with using resources. Thorough accounting of these costs can enable better decisions regarding resource usage and waste prevention. When the total cost of using resources is accounted for, changes in processes, methods, and materials can be more easily justified. The following case studies demonstrate how characterizing the full cost of wasted resources can lead to effective measures for preventing waste.

### Case study: The true cost of using water at an automotive assembly plant

Misunderstanding the true cost of utilizing raw materials such as water can frequently result in poor management choices regarding how the resources are used. Metal-finishing operations are notorious for using large quantities of water in their processes. The common perception is that "water is cheap," so it can be used liberally to ensure that workpieces are adequately cleaned, rinsed, and coated. While it may be true that the actual purchase of the water itself is relatively cheap, the cost of using the water within industrial processes may be considerably more expensive.

A process improvement study, performed by the University of Illinois (Lindsey, 2007) on the paint process line at a major automotive assembly plant, compared costs for purchasing water to costs associated with using it in various processes that prepare vehicles for painting. At the beginning of the project, the automaker perceived its water costs to be only US$2.20 per 1,000 gallons (the cost to purchase water from the city). At this low cost, water was used liberally throughout the plant to ensure adequate quality of cleaning and coating processes. Consequently, conservation measures were difficult to justify from an economic standpoint. However, when the process was broken down on a step-by-step basis and all costs associated with using the water were considered, it was concluded that the true cost of using the water was much higher. When the costs of process chemicals, energy, water purification measures, and waste-water treatment were considered, the total cost of using water increased the average cost to US$80 per 1,000 gallons (a 36-fold increase).

Figure 10.4 shows the breakdown of water use costs on a stage-by-stage basis. As shown, some stages were considerably more

**FIGURE 10.4 Costs of using water on an automotive paint line**

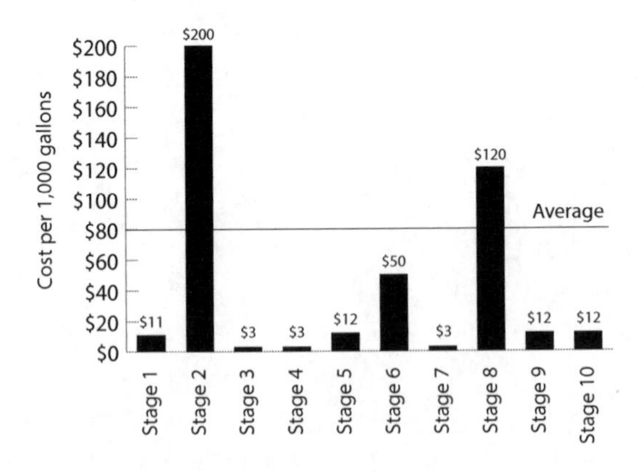


Note: $80 per 1,000 gallons is a weighted average that, in addition to cost, takes into account the different volumes of water used in each stage of the process.

expensive than others. In general, rinsing stages that used municipal water (Stages 3, 4, and 7) were the cheapest while the most expensive step was the degreasing (cleaning) stage (Stage 2) costing US$200 per 1,000 gallons. An analysis of the total costs associated with Stage 2 indicated that process chemicals (for degreasing the vehicles) comprised the vast majority of process inputs (82%) in this stage followed by waste-water treatment (9%), deionized water (5%), and heat energy (5%).

Prior to performing this analysis, the plant had been using about 90 million gallons of water annually because it perceived that water was cheap and conservation measures could not be justified economically. It estimated that purchasing water from the city cost it about US$200,000 per year. The results of the assessment showed that using this quantity of water was actually costing over US$7 million per year. Within a year of the assessment, numerous conservation measures had been justified and implemented. Water usage was reduced by nearly a third, resulting in cost savings of over US$2 million per year.

**Case study: Justification of reusable containers at a manufacturing facility**

Results of a waste audit conducted at a manufacturer of large machinery indicated that wood waste comprised its largest waste-stream based on both volume and cost. Analysis of the sources of wood waste, by the company's supplier of waste management services, revealed that the waste came mostly from wood packaging associated with incoming components from suppliers such as engines, transmissions, and pumps.

Initial reactions to the wood waste problem were focused on reactive measures such as finding low-cost disposal methods. Additionally, various options for recycling the wood waste were considered, including shipping it to a producer of pelletized wood fuels or shipping it to an institution with a boiler that could use wood for fuel. Unfortunately, these options were not economically feasible because the potential users of the wood were not willing to pay a reasonable price for the material or the cost of shipping. A suggestion was made to consider reusable containers for large components instead of building a custom wood box for each item and scrapping the wood after a single use. The recommended containers were collapsible such that six of them could be shipped back to the supplier in the same amount of space occupied by a loaded container.

Initially, economic analysis of the reusable container option indicated that it was not feasible because the cost of purchasing and using reusable containers was about twice the cost of managing the wood waste. In fact, this option had been considered several times in the past but was always met with the same objection. However, additional costs associated with managing the wood waste had not been considered. For instance, the cost of purchasing the wood and the labor necessary to fabricate the shipping containers were not accounted for in the assessment. The manufacturer's accountants objected to including these costs in the economic assessment because they reasoned that the shipping containers were paid for by the supplier who used an independent contractor to fabricate them. Follow-up discussions with the supplier confirmed that the cost of the materials and labor for fabricating the containers was passed on to the manufacturer.

When the total costs for fabricating each container from scratch were added to the cost of disposing of the associated wood waste, the economics changed substantially. In fact, the manufacturer ultimately concluded that the investment required to switch from custom fabricated packaging to reusable containers would pay itself back in the first year. The savings would continue until the reusable containers were projected to wear out (estimated to be about six years), at which time they would be returned to the supplier for refurbishment/recycling.

These case studies demonstrate the importance of collecting relevant cost data associated with activities for the entire system, including suppliers, employees, and customers. The full cost of using raw materials such as water and packaging – not just the cost of purchasing the raw materials – can provide the necessary incentives and justification for making changes. To make sound decisions regarding waste prevention strategies and priorities, it is necessary to establish a thorough understanding of the sources of waste and the full costs associated with wasteful practices. When these aspects are fully understood, measures can usually be justified that can address the source of wastefulness and improve an organization's effectiveness.

# 11

# Sustainability principle #2: Improve quality

> Hang in there. Trust that those winds are blowing away what's not needed while making you stronger.
>
> Unknown

Waste is essentially a defect, an outcome of deficient processes and/or products. Thinking of waste as a defect applies equally well to sustainability performance and conventional views of quality performance. If products are made that, at the end of their useful life, have no value and require disposal, then the product is deficient. Likewise, if processes used to make a product generate by-products that have no value and require disposal, then the process is deficient. Both quality and sustainability are improved when deficiencies in processes and products are corrected to prevent waste. Similarly, improvements in safety, durability, efficiency, and productivity also prevent waste, and contribute to enhancing both quality and sustainability performance.

Deficiencies associated with the design, fabrication, and operation of processes and products are frequently responsible for wastefulness. These deficiencies can negatively affect the quality, safety, productivity, efficiency, and effectiveness of all types of operations. Reducing wastefulness is a common mantra for virtually every type of quality improvement program currently used in the business world. Consequently, eliminating defects and waste through quality improvement measures can be an effective approach to improving sustainability performance. From the time of the first industrial revolution until World War II, conventional business philosophy believed that quality and productivity were at odds with one another. Figure 11.1 shows how most business management experts perceived the relationship between productivity and quality prior to World War II.

Managers believed that measures undertaken to improve the productivity of operations would have a negative impact on quality. Conversely, measures undertaken to improve quality would have a

**Figure 11.1 Pre-World War II perspective on quality and productivity**

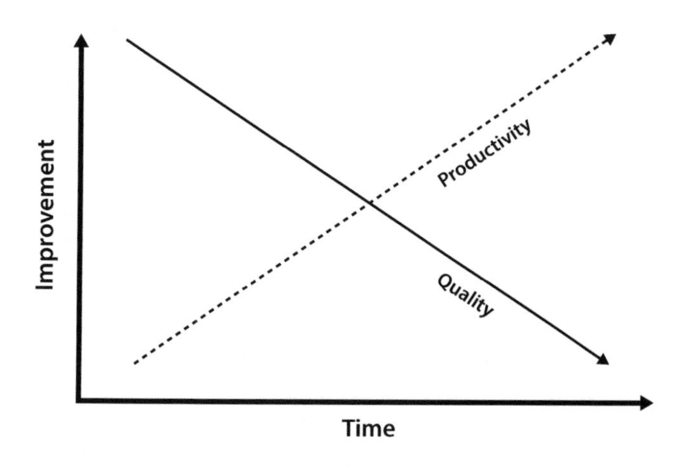

negative impact on productivity. This train of thought was proven specious after World War II when innovative quality gurus, such as W. Edwards Deming, demonstrated that quality and productivity could be improved simultaneously through the *prevention of defects and wastes* (Deming, 1982). This transformational belief was first introduced in post-World War II Japan after American industrial leaders scoffed at the concept. Japanese industry had tremendous success embedding this philosophy in its approach to quality and it is now a central focus of virtually every quality assurance program used by today's business community. Figure 11.2 shows how most business management experts currently view the relationship between productivity and quality.

In many ways, efforts focused on improving sustainability performance are an extension of this philosophy. If we can simultaneously improve productivity and quality through the prevention of defects and wastes, then we can also improve sustainability and quality of life through practices that prevent hazards and waste.

FIGURE 11.2 **Post-World War II perspective on quality and productivity**

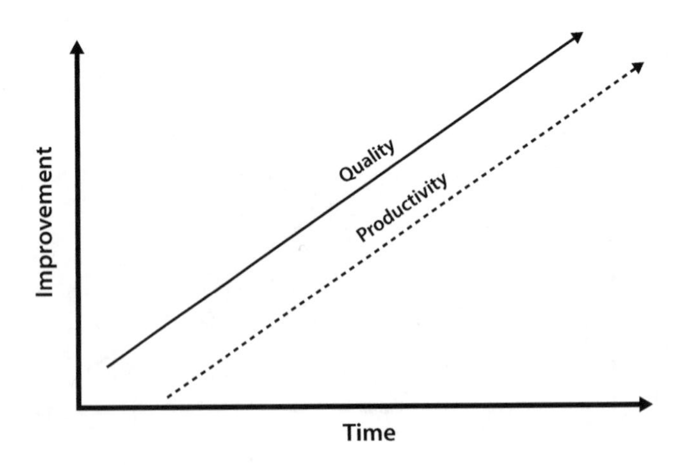

The key to extending this train of thought is to consider the concept of waste prevention more broadly. Seven deadly wastes receive considerable attention in quality assurance programs such as "lean" or "Six Sigma." Emphasis is placed on preventing these wastes because they can greatly reduce an organization's efficiency, productivity. and quality (Womack and Jones, 2003). The seven deadly wastes are:

1. Over-production
2. Waiting
3. Transport
4. Processing
5. Inventory
6. Motion
7. Repair/rejects

The automotive industry was the first sector to focus on improving quality and productivity through the prevention of waste and defects. Today, thousands of organizations across both the private and public sectors have adopted quality systems that are aligned with this approach. Given that waste prevention is a key principle for achieving excellence in both quality assurance and sustainability, synergistic opportunities exist to leverage existing quality systems to drive sustainability performance. This strategy offers considerable advantages because it does not require the establishment of major new initiatives and programs, and enables organizations to emphasize that sustainability is compatible with existing culture. However, overemphasizing the parallels between quality and sustainability can also be problematic because some individuals may conclude that "we don't need to focus on sustainability because we have it covered through our quality system." Quality assurance programs are not usually developed with sustainability in mind, and some important enhancements may be required to ensure compatibility between the two systems.

Although the principles that drive both quality and sustainability performance place considerable emphasis on preventing waste, sustainability excellence requires that waste be considered more broadly. Therefore, expanding the list of seven deadly wastes to also include wastes that contribute to constraint headwinds is a critical step for ensuring strong alignment. Examples of constraint wastes include accidents, human potential, materials, energy, water, and land. Incorporating the prevention of constraint wastes into existing quality assurance programs can facilitate the integration of quality and sustainability initiatives. Considering the additional wastes can also drive monitoring and measurement of additional material aspects that affect the long-term sustainability of organizations. This additional information is frequently critical for developing an effective long-term strategy.

As organizations continuously pursue opportunities to improve quality, they can simultaneously identify opportunities that will improve their long-term sustainability. Most organizations focus on quality with respect to the performance of their products and the processes they use to make them. Customers (both internal and external) who are affected by the performance of the products and processes ultimately define, assess, and judge quality. Many aspects of sustainability performance can be tracked and managed using the same types of methods and measurements used to assess conventional quality aspects such as defects, longevity, efficiency, productivity, and waste.

## Improving sustainability with better quality in processes and products

Sustainability principles need to be considered early and often during efforts to improve existing processes and products as well as during the development of new versions. Process improvements can improve sustainability performance through increases in efficiency and productivity, and prevention of all types of waste, emission, defect, hazard and accident. Notable leaders such as Toyota, Ford, and Honeywell have recognized the connection between quality and sustainability and have woven sustainability principles into their quality systems. Better process controls, procedures, training, materials, facilities, and equipment can all improve processes to achieve considerable sustainability benefits for an organization's operations. However, to be sustainability leaders, organizations must extend their considerations to encompass the entire value chain and life-cycle of their operations and products. This includes design, supply chain, production, customer utilization, and end-of-life options for disposal, recycling, remanufacturing, and redistribution.

For most products and processes, the vast majority of life-cycle impacts are determined during the design stage, when decisions are made regarding key aspects such as materials selection, production methods, maintenance protocols, and energy sources. Designing products to maximize life-cycle benefits while minimizing the resources required for production and utilization is paramount. Products that achieve better efficiency, productivity, and longevity improve sustainability performance because they will require fewer resources to produce and use them. Designing products to be made from materials that are abundant (renewable if feasible), and can be obtained from regions and sources that are politically stable and socially responsible, helps minimize risk and ensures that people and the environment are not exploited. The design stage is also the point where choices are made for energy sources to produce and utilize products, which has a huge effect on emissions. Finally, the design stage includes decisions that will determine options for managing the product at the end of its useful life. Designing for disassembly, refurbishment, remanufacturing, and recycling can greatly extend longevity and perhaps even ensure that the product can be used for multiple life-cycles.

## Microcosms of sustainability: safety and quality

> Life's roughest storms prove the strength of our anchors.
>
> Unknown

Safety and quality are microcosms of the wide spectrum of aspects, opportunities, and decisions that organizations focus on to improve overall sustainability performance. Safety can be directly connected to the social dimension of sustainability while quality can be directly connected to the economic dimension. However, to fully understand the relationships between safety, quality, and overall sustainability performance, safety and quality must be considered more comprehensively than has historically been the case. This can be accomplished by considering how both safety and quality are inextricably linked to all three sustainability dimensions (economic, social, and environmental). The following descriptions explain how safety and quality are integral components of a more sustainable organization.

### The safety aspect of sustainability performance

All responsible leaders are committed to ensuring the safety of their people and their organizations' assets. They will regularly invest considerable time and funds in training, equipment, facilities, personnel, and other assets needed to ensure that their organizations operate safely. The ROI on these investments is virtually impossible to calculate because the benefits accrue in the form of outcomes that never happen. Leaders are aware that cutting back on safety commitments may produce some short-term gains but they recognize that compromising safety will catch up with them in the long term. Nothing is potentially more wasteful than an accident, particularly if someone is killed or injured. Therefore, responsible leaders invest aggressively to minimize risks and ensure the safety of their people and their operations. They recognize that operating a safe organization is good for business.

Safety is obviously a key component of what is required to be a socially responsible organization. It is also strongly connected to the environmental and economic dimensions of sustainability. In addition to directly impacting affected employees, accidents can have negative consequences that extend well beyond an organization's boundaries. Many types of accidents (e.g., explosions, fires, collisions, and spills) can impact citizens and community infrastructure in the vicinity of incidents. Accidents can also impact all types of environmental resources, including forests, rivers, lakes, wildlife, air quality, and water quality. The financial impacts associated with accidents can be extraordinary as costs are incurred both for remediating damaged assets and for shutting down operations while damages are repaired, incidents are investigated, and lawsuits are dealt with. Additionally, an organization's reputation can be damaged along with its license to operate in the future.

## The quality aspect of sustainability performance

Quality is another dimension of an organization's effectiveness that can be difficult to quantify in terms of short-term economic value, but leaders recognize its importance and invest considerable resources in quality assurance. Leaders are aware that reducing their emphasis on quality may increase productivity and profitability in the short term. However, just as accidents can be extremely costly, so can quality failures that lead to:

- Customer complaints
- Waste associated with rejects, scrap, and rework
- Lost customers
- Lost reputation

Therefore, responsible leaders make significant investments to ensure and improve quality, and prevent the costly waste that is usually associated with quality failures. They recognize that an organization's commitment to quality is good for business.

Most organizations are keenly aware of the economic imperative for providing quality products and services to their customers. This is one of the principal means by which organizations deliver value. However, quality is also closely connected to the social and environmental dimensions of sustainability. Quality should be considered broadly to include the quality of the communities where we work and live, and the quality of the environment.

## Considering quality more broadly

As described above, sustainability performance is well aligned with other quality aspects that organizations commonly use to assess and manage performance. However, this alignment does not mean that, just because an organization has a strong quality program in place, it will also exhibit strong sustainability performance. Some significant differences exist between how quality has historically been managed and what is required to embed sustainability in a comprehensive approach to quality (or, depending on your perspective, what is required to embed quality in a comprehensive approach to sustainability). In addition to considering the quality of processes and products as they relate to traditional customers, sustainability excellence requires consideration of additional aspects of quality and how they can affect a much broader customer/stakeholder base. Some examples of additional quality aspects that affect sustainability performance are:

- **Quality of life:** A healthy quality of life for employees and other stakeholders is important for an organization's enduring success. This includes employees and their families, customers, suppliers, and neighbors that may be affected by processes and products. Deteriorating environmental resources, production capabilities, and community resources translate to reduced quality of life for current and future generations, thus wasting some of their potential for achievement. Extreme wastefulness can result in shortages of basic staples that can lead to social unrest and, in some cases, war. When individuals and communities are not able to meet their basic needs, the long-term safeguarding of the environment becomes less of an immediate priority.
- **Quality of communities:** Thriving communities provide many benefits to businesses, markets, and citizens. For instance, communities provide infrastructure, services, housing, employees, and education. They also may provide markets for products and services. Urban decay reduces the quality of life in communities, reduces property values, and erodes the tax base. Degrading infrastructure reduces opportunities for current and future commerce. When a production

facility (such as a factory) is closed, the land, materials, and infrastructure invested in that facility are wasted. Unsustainable farming and forestry practices reduce future productivity through eroded topsoil and polluted bodies of water, wasting their potential for current and future generations.

- **Quality of personnel:** All individuals can contribute to improving sustainability performance regardless of their role or status. Organizations need to develop their full potential by encouraging thought diversity, healthy debate of ideas, professional development, and recognition. Companies that achieve a strong reputation for sustainability excellence can gain a considerable recruiting advantage with respect to attracting talent.
- **Quality of the environment:** A healthy environment is essential for achieving enduring success. The environment is an important supplier of resources (air, water, land, materials, climate, etc.). It is also an important, and well-represented, customer/stakeholder that can be affected by the consequences of wasteful practices. Failure to protect the environment places an organization's future at considerable risk from economic, regulatory, and community relations perspectives. Air, water, and land pollution can cause additional degradation downwind and downstream. Development of greenfield sites can damage multiple types of resources, including cropland, prairie, forests, and wetlands. Deforestation of sensitive ecosystems and conversion of the land to agriculture results in loss of species diversity, soil erosion, and increased run-off. Over-use of national parks and wilderness areas results in degradation of some of our most beloved resources, wasting their potential use by current and future generations.

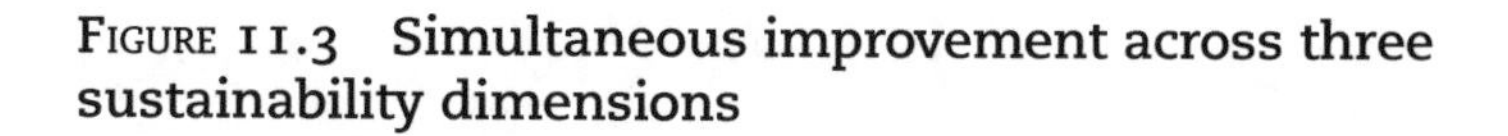

**FIGURE 11.3 Simultaneous improvement across three sustainability dimensions**

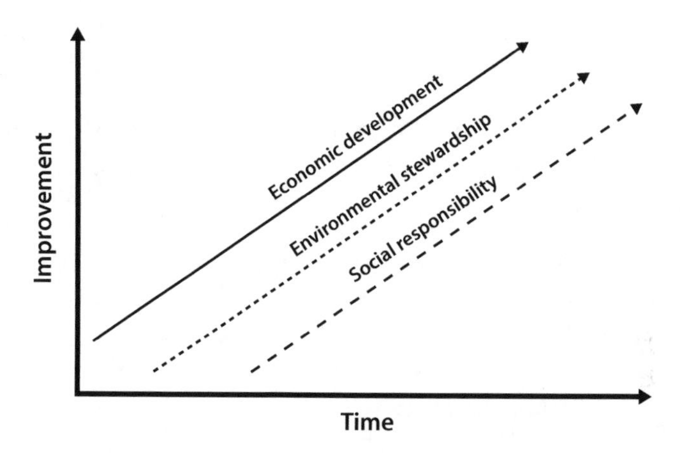


Figure 11.3 illustrates how applying these additional aspects of quality to diverse stakeholders can drive improvement across the economic, social, and environmental dimensions of sustainability.

The following case study demonstrates the importance of understanding how sustainability and quality can be aligned to improve safety and overall effectiveness.

## Case study: Improving sustainability by improving quality – the safe chemistry industry

The chemical industry is a great example of a sector that is undergoing a transformation toward improved sustainability. Chemical companies of all sizes are demonstrating improvements throughout their value chains by incorporating "safer" chemistry (sometimes referred to as "green" chemistry) into their processes and products. Safer chemistry can include:

- Reducing the use of hazardous chemicals and the generation of hazardous wastes
- Reducing the human health and environmental impacts of processes and products
- Creating safer chemical products, such as bio-based chemicals and materials, renewable feedstocks, and less-toxic chemical formulations

The transformation toward safer chemicals is the chemical industry's response to a long history of incidents that have negatively impacted people and community/environmental assets. Chemical spills and remediation liabilities have grown to be huge challenges to the long-term sustainability of the chemical industry and the customers that use chemical products. In 2012, 27,500 toxic chemical spills were recorded globally that were associated with 1,000 deaths (UNEP, 2014). Spills frequently require expensive clean-up and remediation measures, as illustrated by outstanding remediation liabilities of US$722 million for Dow (U.S. Securities and Exchange Commission, 2015) and US$478 million for DuPont (U.S. Securities and Exchange Commission, 2014). The global liability associated with private industry site clean-up is about US$9.7 billion (American Sustainable Business Council, 2015).

Chemical industry customers (including consumers, institutions, large retailers, and public agencies) are demanding safer chemicals from formulators and distributors. The trend is being driven by a variety of factors, including risk mitigation, expanding regulation, NGO and shareholder activism, and customer demands. The rapidly expanding market is incentivizing companies to improve chemical safety standards faster than lagging regulations (American Sustainable Business Council, 2015).

Retailers that sell chemical products directly to customers are also demanding safer chemistry from the manufacturers and formulators that supply them. U.S. retailers that sell chemicals from their shelves as regular consumer products now have to deal with those same products as hazardous waste when damaged items are returned by consumers. Failure to manage returned chemical products as hazardous waste in 2011 and 2012 resulted in fines to leading U.S. retailers as follows (American Sustainable Business Council, 2015):

- Walmart in 2011: US$81.6 million
- Target Corp. in 2011: US$22.5 million
- Walgreens Co. in 2012: US$16.6 million
- CVS Pharmacy in 2012: US$13.75 million
- Costco Warehouse in 2012: US$3.6 million

In response to the growing risk and liability, Walmart proactively engaged with scientific experts and industry groups to identify "high priority chemicals" that need to be phased out of the products they sell. Walmart is working with its suppliers to reformulate its products with safer substitutes that will continue to meet customer expectations. As of 2014, Walmart had successfully removed 95% of the targeted "high priority chemicals" from the products it sells in the U.S.A. (Walmart, 2017).

Improvements in chemical safety are occurring rapidly, as are research and development efforts to produce innovations that are safer for people and the environment. Markets are responding favorably to the safer products. The global market for safer chemistry has grown steadily since 2010 and is projected to grow from US$11 billion in 2015 to nearly US$100 billion by 2020. The North American market for safer chemistry is projected to grow from US$3 billion to over US$20 billion during the same period (American Sustainable Business Council, 2015).

Technological advances continue to create opportunities for growing and preserving markets through safer chemicals. The number of patents for "sustainable chemistry" has increased significantly in recent years, suggesting that research and development investment is increasing and companies are viewing safer chemistry as an opportunity for competitive advantage and brand differentiation. Future innovation will likely be concentrated with specialty chemicals, relatively niche chemical formulators, and small businesses, which can serve the growing demand through their nature as test beds for problem-solving and their often mission-driven approach to product and service development (American Sustainable Business Council, 2015).

By considering quality and safety more comprehensively to include their economic, social, and environmental aspects, their connection to sustainability performance can be easily demonstrated. Similar connections can be made with many other aspects of an organization's sustainability performance, including innovation, human resources, strategy, marketing, reputation, brand, and shareholder value. Consequently, sustainability performance can be a primary driver for improving an organization's effectiveness and value, and should be regarded as a core business function that applies to everyone. Just as everyone can contribute to producing quality products and creating a safe work environment, everyone can contribute to working more sustainably. **An organization's commitment to sustainability is good for business.**

12

# Sustainability principle #3: Optimize systems

> The wind and the waves are always on the side of the ablest navigator.
>
> Edmund Gibbon

Since industrialization began, organizations have focused on consolidating and optimizing various aspects of their enterprises, including access to raw materials, product designs, production processes, distribution to customers, and customer value. Optimization of these aspects requires active participation from various units within an organization as well as cooperation from suppliers, customers, communities, and the environment. Efforts to optimize interactions between these aspects tend to occur in isolation without much regard to their effects on upstream and downstream stakeholders. As organizations grow and become more complicated, processes and procedures tend to become more prescriptive and institutionalized. Layer after layer of policies and procedures stack up to increase complexity and bureaucracy. Employees are trained and directed to strictly follow the established procedures and their performance assessment is frequently tied directly to their track record with respect to following those rules and procedures. These factors merge to form complex systems with multiple aspects that interact through dysfunctional exchanges of materials, energy, information, and other resources. The dysfunctional systems lead to a variety of unintended consequences and complex problems that scientists and engineers struggle to address. Scientists respond to the problems by complicating them to a point where they become virtually unsolvable, while engineers tend to simplify problems to a point where they become solvable but the solutions end up not being effective.

By taking a systems-based approach to addressing problems at their root cause, more effective solutions can be identified and implemented to address the sources of dysfunctional interactions and optimize system performance. Optimizing system sustainability can be accomplished by ensuring that each component is appropriate, functions correctly, and is properly integrated with other aspects. It would be ludicrous for an auto-maker to design and optimize individual components of an automobile (engine, transmission, steering, wheels, chassis, etc.) without consideration of the other components. To do so would inevitably produce a vehicle that would not function well as a system. In fact, the notion is ridiculous because, in this example, an automotive company is in control of all components of the system and can ensure that the components are designed, fabricated, and integrated to form an efficiently functioning system. In most man-made systems, however, the design and management of the individual components is not controlled by a single entity. In these cases, the individual components tend to be optimized in isolation without much regard for how they will impact other components. Consequently, many interactions between components tend to be dysfunctional, resulting in deficiencies in products and processes and considerable wastefulness.

## Examples of perfectly optimized systems

It has long been understood that Mother Nature provides living, breathing examples of the most efficient systems on the planet – in the form of ecosystems. As shown in Figure 12.1, energy is inputted directly from the cleanest of all sources, the sun, and is collected by producers (plants). The plants use the solar energy to drive photosynthesis as they convert $CO_2$ and other nutrients into biomass that serves as a source of energy and habitat for consumers (animals). The by-products from plants and animals are processed by Nature's recyclers, the decomposers (e.g., earthworms and microorganisms). These organisms break down the by-products into forms that can

FIGURE 12.1 A perfect system

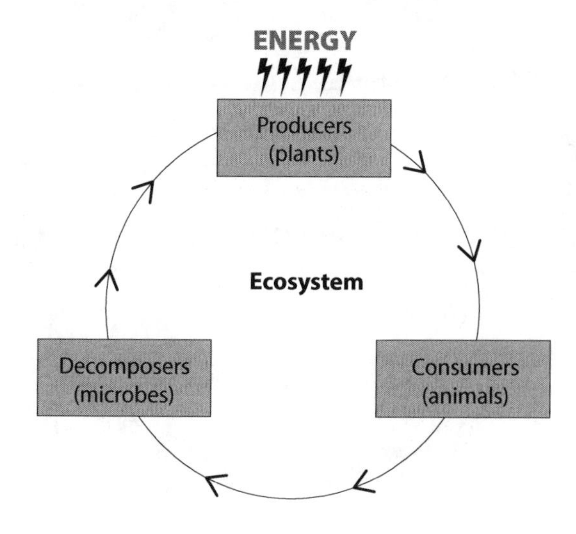

again be utilized by plants and animals. Energy and materials continuously move from producers to consumers to decomposers through a circular flow that exhibits elegant balance and efficiency. Nothing is wasted or degraded regardless of the ecosystem in question (forest, desert, ocean, etc.). The interactions between the components of natural systems are perfectly optimized in terms of sustainability.

Man-made systems are not perfect. As described by Deming (1982) and portrayed in Figure 12.2, man-made systems are comprised of interactions between materials, energy, people, machines, methods, surroundings (environment, market conditions, political climate, etc.), and the way they are managed. Managers and engineers tend to focus primarily on how to design, operate, and improve the specific components of systems for which they are responsible. Consequently, individual components of man-made

FIGURE 12.2 **Man-made industrial systems**

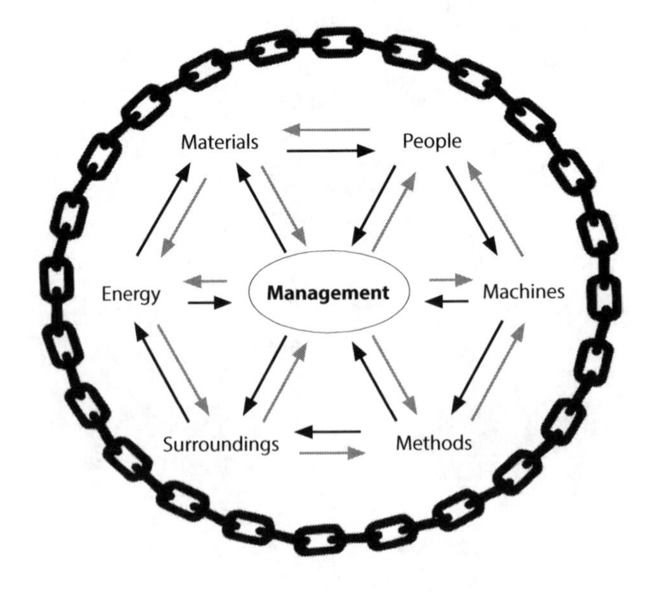

systems, such as processes, products, projects, and parts, tend to be optimized in isolation.

Today's industrial systems tend to be complex and technical, and the various aspects regularly interact in ways that are dysfunctional. Efforts to improve efficiency in one component may produce benefits for that aspect but may also create unintended consequences for other internal aspects as well as interactions with suppliers, customers, communities, and the environment.

Dysfunction is frequently created during the design stage where the vast majority of a product's life-cycle impacts are established. For instance, an engineer responsible for designing a machine may determine that spot welding a part in place is more cost-effective than bolting it into place. Consequently, he specifies spot welding in the design specifications. He may even receive recognition for developing procedures that reduce costs and improve the bottom line. Unfortunately, this specification may actually add to the total costs of owning and operating the machine and reduce its life-cycle benefits. Welds are meant to provide permanent fastening whereas bolted items can be easily removed and replaced. The welded part may prove to be problematic for customers that need to service or replace the part periodically. It also may prevent remanufacturing or rebuilding the worn part. In some cases, the entire machine may be scrapped when the welded part fails because it can't be efficiently serviced or repaired. A small additional investment in measures that enable bolting the part in place would slightly raise the initial cost but would greatly increase the product's longevity and reduce the life-cycle costs and resources needed to own and operate the machine.

## Engineering the chain ... not just the links

System optimization requires thinking beyond products and the processes used to make them, to consider end-to-end solutions and actual customer needs. Most conventional business models focus on selling as many items as possible to as many customers as possible. In contrast, systems-based models focus on providing enduring solutions to customer needs. This includes opportunities for leasing and selling function and performance as opposed to selling "stuff." Systems-based models also include incorporating the social and environmental costs and benefits, such as:

- Implications for health and wellbeing
- How to reuse and recycle components and materials
- Using renewable sources of materials and energy
- Enabling business partners to increase their efficiency and productivity with resources

Consideration for all the benefits of more sustainable systems can often lead to the identification of new opportunities and revenue streams that can expand and diversify markets.

System optimization is best achieved by engineering the entire value chain to ensure that individual components function to the benefit of the overall system. Performance gains achieved by improving one aspect of the system should not be accomplished to the detriment of other parts. To achieve this, planners and designers need to communicate regularly and collaborate to ensure that their plans and designs do not cause negative impacts to the efforts of others. This includes production operations, the supply chain, distributors, recyclers, regulators, and customers. Designers that are particularly skilled in system optimization often refer to this practice as **"engineering the chain ... not just the links."**

Products should be designed to maximize the life-cycle benefits of resources while minimizing consumption, cost, and negative impacts. This includes designing them to be energy efficient, and facilitating regular maintenance and convenient replacement of components that wear out at faster rates. In some instances, low-value sacrificial components can be incorporated to protect higher-value components and to facilitate maintenance. To keep pace with rapid technology advances, products need to be designed so they are upgradable. Materials should be selected to ensure that worn components can be mined for use in remanufacturing, rebuilding, and recycling. As organizations optimize their systems, considerable benefits accrue, but not all stakeholders benefit equally. Some stakeholders gain benefits while others are forced to sacrifice. Some constituents are usually required to invest and/or sacrifice more resources than others. Additionally, the benefits of system optimization are usually compelling when the system is considered in total, but they are usually not distributed evenly among the various stakeholders.

This dynamic can create considerable conflict unless strong leadership is exercised and measures are taken whenever possible to allocate sacrifices and benefits equitably. For instance, many organizations have historically allocated the costs associated with managing industrial wastes to the environmental, health, and safety (EHS) department because they are responsible for ensuring regulatory compliance. However, the EHS department does not actually generate much waste and, therefore, can do very little to prevent waste generation. Additionally, the vendor that the EHS unit hires to manage and dispose of the waste is usually compensated based on the quantities and types of waste that they manage. Therefore, they have no monetary incentive to help their customer reduce waste generation.

Most industrial waste is generated from a combination of maintenance, production, and other operational activities. Since these units are not forced to pay for the costs of waste management and disposal, they often perceive that they have little incentive for changing practices to reduce wastefulness. A more optimized system would allocate the costs of managing waste to the units that are responsible for its generation. This change provides incentives for those in the best position to make changes to find opportunities for preventing waste through improved efficiency and productivity measures. Additionally, the waste management vendor's contract could be modified to include incentives for reducing waste through implementation of prevention and recycling measures. By aligning the interests of all the stakeholders, collaboration and cooperation is improved and optimization measures become easier to implement.

## New business models focus on function and performance

Many opportunities exist for optimizing interactions with suppliers and customers that can produce considerable benefits. System optimization gains can often be accomplished by improving interactions with key suppliers, customers, and strategic partners. Specifications regarding product designs and the selection of materials, processing methods, fuels, packaging, shipping, and so on frequently have direct impacts on the sustainability performance of suppliers and customers. Business models often institutionalize many of the interactions between suppliers, producers, and customers. In instances where suppliers primarily sell goods such as materials, chemicals, food, energy, machines, and devices, the supplier's interests are not well aligned with the interests of their customers. Suppliers can frequently generate more short-term revenue by convincing their customers to purchase, use, and discard more products. This model tends to encourage suppliers to promote excessive consumption and waste generation by their customers. These types of interaction are dysfunctional because they encourage wastefulness and reduce sustainability. New business models have emerged in recent years that align the interests of suppliers and customers such that the suppliers generate more revenue by supplying less product.

Business models that are dependent on selling or renting "stuff" are becoming increasingly vulnerable to competing models that use technology to replace stuff with information. Customers typically need less "stuff" as systems improve and waste is eliminated; however, they still need function and performance. For example, movie rental stores such as Blockbuster were decimated as online and on-demand movie rental systems such as Netflix and Amazon became mainstream. Blockbuster chose not to evolve with the changing technology because it perceived that it was in the business of renting and selling tapes and disks when, in reality, it was in the entertainment business. Customers didn't want to rent tapes or disks – they only wanted to watch movies and programs. Netflix and Amazon leveraged innovation to optimize an entertainment system that provided the function and performance that customers desired with considerably less hassle and waste. Customers no longer needed to travel to a movie store to obtain a movie, then make another trip to return it; and they no longer had to pay late charges for failing to return the rented items on time. Customers experienced multiple benefits from the transformation, including time savings, reduced fuel consumption, cost savings, and reduced travel risk.

Business models focused on selling function and performance instead of selling "stuff" have become increasingly popular in recent years. These performance-based models are structured to align the interests of both suppliers and customers such that both entities benefit from improved efficiency and reduced wastefulness. Customers are able to optimize multiple aspects of their operations, including some that are not core to their business, without developing extensive in-house capabilities and expertise. They structure agreements with contractors and suppliers to ensure they provide the deep expertise and capabilities needed to drive sustainability innovation in ancillary aspects of their operations. This frees up customers to focus on improving core aspects of their operations.

These business models can also help preserve margins for suppliers and contractors that might otherwise be in jeopardy in cases where system optimization reduces the customer's need for "stuff." Several examples are provided below that describe how transformational business models have disrupted markets in recent years. In each case, innovative practices have been implemented that changed the focus of interactions from selling "stuff" to selling function and performance. Considerable gains in system optimization and sustainability performance have been achieved in each case.

## Chemical management replaces chemical purchasing

Historically, chemical suppliers have been compensated for the quantities of chemicals that they provide to their customers. Their profits have been tied to chemical volume – the more chemicals sold, the more profit generated. These agreements enable suppliers to benefit from customer wastefulness associated with excessive consumption, inadequate maintenance, and lack of recycling. Systems based on these arrangements are dysfunctional because they frequently lead to excessive costs, risks, and waste management obligations. In the 1980s, a supply chain management strategy commonly referred to as "chemical management" emerged in the automotive industry to address the dysfunctional aspects of the system and align the interests of the suppliers with the customers.

Under a chemical management contract, vendors such as Houghton, HAAS, Quaker, Henkel, Fuchs, and Castrol are no longer compensated primarily on chemical volume. Instead, compensation is tied primarily to services delivered and chemical performance. Most automotive and aerospace companies, along with some other large manufacturers, no longer purchase chemicals such as paints, lubricants, adhesives, and solvents from multiple suppliers. Instead, they engage with a single Tier I supplier in a strategic, long-term relationship where the supplier serves as a "chemical manager." The chemical manager is responsible for supplying and managing all of the chemicals and related services necessary to successfully make a product. Chemical management goes beyond supplying chemicals to include optimizing processes – continuously reducing chemical life-cycle costs, waste, risk, and environmental impact.

The chemical manager is expected to find efficiencies during the course of its contract through measures that reduce waste, improve productivity, and enable recycling. Instead of paying for chemicals, the customer pays the chemical manager for the successful performance of the chemicals either on a cost per unit basis (i.e., price per car) or a cost plus services basis with gain-sharing incentives. In many cases, both the customer and supplier share in the cost savings achieved through implementation of the efficiencies. This model aligns the interests of the supplier and the customer because the supplier actually makes more money by supplying less chemicals. Considerable cost reductions can be achieved through chemical management programs because reductions in chemical use exceeding 50% are common. The savings frequently extend well beyond the cost of chemicals to include benefits from better inventory management, reduced waste, and improvements in productivity and product quality (Chemical Strategies Partnership, 2017).

According to practitioners in the chemical management sector, the current market is estimated at US$1.2–1.5 billion per year and continues to grow as customers focus more on core business and grow more comfortable with their supplier's chemical management capabilities. Great potential exists for expanding this model to incorporate other system components. For example, some chemical managers have begun working with tooling suppliers to optimize system interactions and extend tool life.

## Resource management replaces waste management

Resource management (RM) is a strategic alternative to conventional waste disposal contracting, where waste generators and contractors share financial benefits from "resource efficiency" innovations. RM emphasizes cost-effective resource efficiency through prevention, recycling, and recovery – instead of waste hauling and disposal. RM contracting is based on three premises:

1. Significant cost-effective opportunities exist to reduce waste, boost recycling, and optimize services
2. Contractors will pursue them when offered proper financial incentives
3. Financial incentives to contractors are supported by the savings generated through cost-effective improvements to the customer's system for managing wastes

For example, if contractors identify cost-effective recycling markets for disposed materials or techniques for preventing waste altogether, they receive a portion of the savings resulting from the innovation.

This arrangement enhances the recovery of readily recyclable materials while promoting opportunities to develop new markets for difficult-to-recover materials. As a result, RM promotes a business-driven effort – rather than regulatory initiatives – to make waste prevention a priority.

As a performance-based contract strategy, RM also taps into the expertise of external contractors to bolster waste reduction and recycling through value-added services, such as improved data collection, analysis, and reporting. The key to success in RM contracting is changing the compensation structure. By providing incentives for contractors that reward them for achieving mutually determined goals, the contractors' profitability model is shifted from "haul and dispose more volume" to "minimize waste and manage resources better." This aligns the interests of the customers that generate waste with those of the RM contractor, by constraining disposal compensation and providing opportunities for both the contractor and the customer to profit from resource efficiency innovations. This arrangement enhances recovery of readily recyclable materials such as cardboard, wood, and plastic while promoting market development opportunities for materials that are more difficult to recycle, such as paint sludge and solvents.

GM adopted RM in response to an internal corporate waste reduction goal and the recognition that existing hauling and disposal contracts produced limited and uncoordinated resource recovery across its more than 70 North American facilities. GM's objective, in executing RM contracts, was to "provide a systems approach to resource efficiency that motivates cost reduction and conservation of plant resources" (EPA, 2017e). A number of diverse organizations are adopting similar best management contracting practices. These include municipalities such as the city of Omaha, Nebraska; corporations such as Kinko's and the Ford Motor

Company; and institutions such as the West Des Moines, Iowa School District (EPA, 2017e).

Waste Management Corporation (WM) began offering RM services in 1997 through its "Sustainability Services" program. The program has grown steadily since then to a current team of 400 engineers, architects, and other specialists. It works directly with customers in multiple sectors to develop and implement an optimized suite of innovations and processes that minimize waste, increase recycling of by-products, and improve compliance performance.

In 2015, WM managed 141 sustainability service programs in the following sectors:

- 7 in metals, pulp and paper
- 25 in refining/energy
- 34 in auto manufacturing
- 5 in pipelines
- 16 in chemical/pharmaceutical
- 54 in manufacturing

These programs helped WM customers save US$12.4 million in 2015 and have saved customers US$165 million since the program's inception through improved efficiencies with materials and energy and reduced needs for waste disposal (Waste Management, 2017).

## Lighting replaces light fixtures and bulbs

Rapidly improving performance associated with light-emitting diode (LED) lighting technology has disrupted the market for light fixtures and bulbs. LED lights last about 50 times longer than a typical incandescent bulb, 20–25 times longer than a typical halogen bulb, and 8–10 times longer than a typical compact fluorescent light. They use 50–70% less electricity, produce very little heat, contain no mercury, and are resistant to shock and vibration. LEDs can operate effectively in extremely cold environments, and they can be programmed and controlled through a variety of electronic devices. LED systems are controllable and adaptable to changing conditions and can help create more comfortable and productive environments for people (Bulbs.com, 2017).

The ability of LEDs to accept digital signals can be leveraged beyond simple illumination to make them a key component of the "internet of things." LED "smart lighting" systems can help provide a comprehensive view of what is happening in a facility. Data gathered from a smart lighting system can determine that a conference room is underutilized and facilitate better utilization of that space (Weinschenk, 2016). A sensor can track people flow through buildings and provide critical input regarding space utilization and resource allocation. Temperature sensors can determine if the heat or air-conditioning is being turned on too early in the morning (Weinschenk, 2016).

Philips Lighting began manufacturing and distributing light bulbs and fixtures in 1891. Historically, Philips has helped customers meet their lighting needs by supplying them with various fixtures and bulbs. Regular bulb replacement provided a recurring revenue stream when bulb life was limited. However, bulbs are no longer a commodity now that LED technology has disrupted the lighting market. If Philips simply added LED bulbs to its product line using its conventional historical business model, it would undermine a large portion of its sales associated with bulb replacement (Fleming and Zils, 2014).

Philips has responded to the LED technology disruption in the lighting market by offering its business and municipal customers a connected system approach where it sells lighting performance instead of fixtures and bulbs. Customers pay Philips for the light and Philips supplies the technology and manages the risk and investment. In many cases, Philips also takes obsolete equipment back when it becomes obsolete and either upgrades it for reuse or recycles the materials. Light could increasingly come to resemble a utility, like water or gas, whereby a company such as Philips builds the infrastructure to provide it where it's needed, and then gets paid for it (Gertner, 2014).

## Energy as a service replaces energy as a commodity

For decades, electric power has been generated by utilities and distributed as a commodity to customers over a power grid. However, technology disruptions are causing the energy industry to become more complex with respect to how energy is generated, distributed, managed, and regulated. Additionally, many opportunities exist to improve efficiency in buildings and production processes. Staying current with the latest technology developments can be challenging for many organizations.

To ensure that they gain the benefits of emerging technology without taking on the capital investment and complexity burden that often comes with it, some organizations are turning to "energy as a service" (or "energy performance") contracts with service providers. Instead of selling electricity, service providers sell bundled services that most utilities do not. The "energy as a service" business model provides customers access to technology options regarding how they implement, optimize, and manage their energy infrastructure. This ensures that they experience maximum benefits with the lowest possible energy use, cost, and impacts.

Energy service providers identify, integrate, and manage appropriate energy technologies such as on-site generation, combined heat and power, renewable energy, energy storage, sub-metering, and grid integration. They develop, design, build, manage, and finance projects that save energy, reduce energy costs, and decrease operations and maintenance costs at their customers' facilities. They also act as project developers for a comprehensive range of energy conservation measures and can assume the technical and performance risks associated with a project. When an energy service provider implements a project, the company's compensation is directly linked to the customer's actual energy cost savings.

The model provides customers choices with respect to:

- How energy is generated and the environmental impact of that generation
- Who owns the energy generation assets (the user, the supplier or a third party)
- The level of dependence on the grid and/or distributed energy resources
- The degree of redundancy and resiliency built into their energy capabilities
- The pricing and payment mechanisms
- Bill consolidation across multiple locations

Edison International recently launched Edison Energy to provide energy as a service to customers. It identifies solutions and design customized systems to meet customer needs and goals in the most cost-effective and sustainable fashion, while managing their exposure to energy risks (Edison Energy, 2016).

Opportunities for expansion are unlimited as service providers take advantage of smart meter capabilities that facilitate diverse service offerings in smart home technology such as home security or monitoring of the elderly. Expansion into providing other resources is also likely and some service providers are considering expansion into water and waste management. The National

Association of Energy Service Companies maintains a list of energy service providers (NAESCO, 2017).

Like many businesses, chemical manufacturers, auto-makers, energy utilities, and so on commonly talk about headwinds constraining their businesses. Do you suppose that the companies engaged in the function and performance models described above perceive that constraint headwinds are an obstacle or an opportunity for market expansion? Many additional examples of business models driven by function and performance are emerging in today's markets. They range from Interface Carpet selling a service that keeps customers' floors covered, to GE selling jet engine performance by the hour of use. Regardless of the application, these models share a common theme – the interests of the supplier and customer are aligned to create interactions that are inherently less wasteful and more fully optimized.

## A growing natural resource: information

While the quality and availability of most natural resources continue to decline, one prominent exception to this rule has emerged in recent years: information. Fortunately, the growth of the information resource can play a critical role in mitigating declines in other resources. The industrial world is increasingly merging with the digital world as technology advances – associated with sensors, data management, and communication – have been leveraged to connect all types of device and establish an "internet of things." These innovations have dramatically improved information availability and flow to create system optimization opportunities that were not conceivable just a few years ago. The network of devices connected by communications technologies is growing exponentially to create systems that can monitor, collect, exchange, analyze, and use data to effectively optimize systems. By combining machine-to-machine communication, industrial big data analytics, technology, and cybersecurity capabilities, the internet of things is driving unprecedented levels of efficiency, productivity, and performance. Industrial operations in diverse sectors such as utilities, agriculture, oil and gas, manufacturing, healthcare, aviation, and many others are experiencing transformative operational and sustainability benefits.

As innovators continue to use the internet of things to optimize systems, some longstanding business models become vulnerable to new versions that are less wasteful, more efficient, and more productive. Disruptive innovations that can facilitate system optimization are becoming increasingly common. As the internet of things generates additional market disruptions, many, if not most, sectors will be transformed through improved system optimization. The following case study demonstrates how system optimization can be improved through the internet of things.

**Case study: Information technology and mobility**

Automobiles are frequently one of the most expensive assets that people own. However, automobiles sit idle about 96% of the time. People have, historically, tolerated the poor utilization rate in exchange for ensuring they have convenient and reliable access to transportation. This wasteful and dysfunctional system has led to heavy traffic, long commutes, and negative environmental impacts. Better vehicle utilization could reduce these impacts considerably, such as the need for parking and the land requirements associated with it. Parking accounts for as much as 24% of the area comprising American cities. People looking for parking account for 30% of miles driven in urban business districts (Bertoncello and Week, 2015).

Multiple innovators are now focusing on alternative modes of transportation, particularly through various connected driving options. The mobility sector is evolving from a system based on individual car ownership toward a variety of customer choices and service levels connected by systems optimized by improved access to data and communication. E-hailing, car sharing and autonomous driving are all in various stages of impacting the existing transportation market, and will continue to do so for decades to come. They can help increase car utilization and reduce the number of cars needed by the growing population (The Economist, 2015b).

### *E-hailing*

Customers in need of a ride no longer have to call, schedule, or hail taxis on busy street corners to gain access to transportation. The taxi industry is experiencing unprecedented competition from ride-sharing systems such as Uber and Lyft. Customers can utilize an app on their smartphone to connect them with drivers that can provide rides. They can track the car picking them up on their smartphone and know precisely when it will arrive. Customers pay for the service through mileage-based fees charged to their credit cards that the company keeps on file; a receipt is automatically sent to the customer's email account. Uber and Lyft collect a percentage of each fare while the remainder is distributed to the drivers.

Uber has developed a relatively efficient system for drivers to become subcontractors. The drivers, their documentation, and their vehicles are scrutinized and approved or declined. The day-to-day relationship between Uber and their drivers is managed almost entirely though the app. The customer preregistration system facilitates the payment process and drivers are not required to process transactions or carry cash in the car. Drivers are rated by passengers and low-scoring drivers can be suspended from the Uber service. The ratings process incentivizes drivers to offer substantial customer service. The taxi industry has argued that Uber engages in unfair competition because it operates outside of the regulatory burden they are forced to comply with (McGregor *et al.*, 2015). If the taxi industry expects to survive the rise of e-hailing services, it will need to shift its focus toward innovation rather than regulation.

E-hailing services provide a variety of economic, social, and environmental benefits. Drivers are independent contractors with a great deal of flexibility and independence. Uber drivers regularly achieve higher capacity utilization rates than taxi drivers in most cities. On average, the capacity utilization rate is 30% higher for Uber drivers than taxi drivers when measured by time, and 50% higher when measured by miles. Since they do not require cars to be hailed, they can more easily pick up customers in less developed neighborhoods (Cramer and Krueger, 2015). Founded in 2009, Uber offers services in over 66 countries and 449 cities worldwide (Uber, 2017). Expansion of this model is inevitable and, in the near future, e-hailing drivers will be used to deliver mail, packages, groceries, and fast food.

### *Car sharing*

Car-sharing services have grown in recent years to offer an economical, convenient, and socially conscious alternative to car ownership. They allow customers to find a car nearby, book it, pay for it, and, in some cases, open it, all with their smartphone. The convenience of the transaction compares favorably against conventional car rental systems where customers stand in line, fill out paperwork, inspect the vehicle, and receive a sales pitch for extras such as insurance. Car-sharing services have been growing by 35% per year in the U.S.A. in recent years, reaching 1.6 million members in 2014. In Germany, car-sharing membership has grown 50% per year since 2010, reaching 1 million people in 2014 (Bouton *et al.*, 2015).

The biggest advantage of car-sharing services is accessibility. Reservations can be made with a convenient app on a smartphone or computer. And unlike traditional rental car companies, which have set hours of operation, pick-up and drop off is available 24/7, 365 days a year, often from a conveniently located parking spot or garage. This eliminates the need to travel to a conventional car rental outlet, wait for an airport shuttle bus to pick you up, or stand in a long line at the rental counter to access a car.

Another equally attractive benefit of a car-sharing service is flexibility. Cars can be rented for a quick errand – as short as 30 minutes – and customers pay only for that time. Traditional rental car companies typically require a minimum one-day rental (ForbesBrandVoice, 2014).

Zipcar, founded in 2000, was an early leader in the car-sharing business. As of June 2015, it offered nearly 10,000 vehicles throughout the U.S.A., Canada, the U.K., Spain, France, Austria, and Turkey. Zipcar was acquired by the Avis Budget Group in 2013 (Naughton, 2013).

In response to the success of Zipcar, multiple other car rental and car manufacturers have also established car-sharing offerings, including:

- Enterprise: CarShare
- Hertz: On Demand
- U-Haul: U Car Share
- Daimler: car2go
- BMW: Drivenow

Ford announced a pilot program in January 2016, where it planned to offer 24-month leases on vehicles to self-organized groups of 3–6 people in Austin, Texas (Korosec, 2016).

While rentals and fleet-based car-sharing services are found predominantly in major metro areas, "peer-to-peer" sharing is without geographic restrictions. These car-sharing platforms connect private car owners with would-be car renters. The peer-to-peer car-sharing giant, Turo, lists cars in multiple U.S. urban, suburban, and rural locales. Turo allows car owners to connect with people in need of a rental car. Car owners set their own rates and approve each rental request, while Turo provides insurance to cover the rental and handles payment. You can list your car either at home, or at a local airport to attract travelers (Kesler, 2015).

### *Autonomous vehicles*

Driverless technologies are becoming increasingly common in vehicles. Some of today's cars can parallel park into tight spaces, maintain safe following distance, and keep vehicles centered in a lane. Some vehicles will even apply the brakes when they sense a collision is imminent. As these innovations become more common, they will increasingly be combined into systems that will be controlled by sophisticated software to establish a comprehensive autopilot. Not only does autonomy show promise in the automotive sector, various other adjacent equipment applications are adopting technology to improve autonomy. For example, in the construction and warehousing sectors, applications are under development for machines such as excavators, forklifts, loaders, and haul trucks.

Autonomous vehicles will likely be safer, cheaper, and more convenient than the vehicles currently on the road. This technology offers the potential for significant social and environmental benefits. Driverless cars cannot drink alcohol, break the speed limit, or be distracted by a text message, so accidents should occur much less often. By optimizing the driving system, autonomous vehicles will reduce travel-based labor, fuel consumption, and associated emissions. They will also create new leisure options for travelers, and improve productivity for commuters. Autonomous vehicles can also potentially increase mobility for the elderly and disabled (The Economist, 2015b). Passengers in self-driving vehicles would be freed up to do other things. Morgan Stanley estimates that the resulting productivity gains would be worth US$1.3 trillion a year in America and US$5.6 trillion worldwide (RobotEnomics, 2014).

A variety of automotive industry suppliers and IT companies are engaged in actively developing technology to facilitate autonomous vehicle deployment:

- Google is developing the software that machines need to think and react smoother and faster than a human driver
- Cisco is developing components that transport data, voice, and video inputs in ways that transform how people connect, communicate, and collaborate
- Delphi is developing affordable ways of reducing the complexity of advanced safety systems
- Continental is developing technology to connect cars via wireless networks to improve real-time traffic and navigation, passenger entertainment features, and safety hazard warnings

Various other companies specializing in data monitoring, communication, and management systems are engaged with the automotive sector to help make autonomous vehicles a reality (Tannert, 2014).

In spite of a variety of legal, regulatory, and insurance issues that challenge the expeditious deployment of autonomous vehicles, some forward-thinking companies continue to aggressively pursue the technology. Since 2009, Google's autonomous vehicle fleet has driven more than 1.5 million miles and caused just one accident. Google estimates that shared, self-driving taxis could eventually achieve utilization rates of more than 75%. Consequently, in the future, a much smaller number of cars could be needed to transport people (Stewart, 2016). Experts estimate that highly automated vehicles will be here by 2020. Ford CEO Mark Fields announced in August 2016 that Ford intends to have fully autonomous vehicles in commercial operation for a ride-hailing or ride-sharing service beginning in 2021 (Fields, 2016).

## The "edge effect": finding innovation opportunities where systems intersect

In mature systems, opportunities for innovation can be difficult to identify because core processes and procedures have been refined and optimized over extended periods. However, productive opportunities frequently exist at system fringes that are given little attention. Exploiting these opportunities frequently requires effective collaboration between partners that have not traditionally worked together. Productive innovation opportunities can often be found along the boundaries of adjacent sectors where interactions can be optimized through effective partnerships.

Anyone who has done much hunting or fishing has likely experienced the productivity of the edge effect. The edge effect is a biological phenomenon that describes how some species of animals tend to thrive along habitat borders resulting in higher populations and greater biodiversity. Consequently, hunting and fishing expeditions tend to be more productive when they focus on areas with edge effects, such as where a pasture meets a forest, where a stream discharges into a river, where a river flows into the sea, and so forth. Just as greater hunting and fishing opportunities tend to exist along the borders of natural systems, opportunities for sustainability innovation tend to occur along the borders of various sectors. Some examples are provided below.

## Waste heat and $CO_2$ for greenhouses

One popular example of how adjacent systems can be connected to produce sustainability benefits involves optimization between the energy, agriculture, and transportation sectors. In northern climates, hundreds of greenhouse operations are now collocated with distributed power generation operations. Waste heat from power generation is recovered and used to heat the greenhouses. This inexpensive source of heat is very valuable to greenhouse operators because their energy costs are usually one of their biggest expenses. In addition to recovering the waste heat, $CO_2$ produced from fuel combustion is also recovered and piped into the greenhouse where it becomes fertilizer that increases the productivity of plants. Additionally, by producing vegetables locally in the greenhouses, millions of miles of truck traffic can often be avoided from shipping in vegetables from distant locations. The resulting systems can be over 90% efficient with respect to energy utilization. In many cases the system can actually be considered carbon negative when compared to conventional systems – even though the power is generated with fossil fuels. Much of the $CO_2$ that is generated becomes sequestered in the plants, plus considerable emissions are avoided by producing plants locally instead of shipping them in from warmer climates.

## Geothermal energy from depleted oil and gas wells: an opportunity ripe for the taking

As described in Chapter 5, renewable geothermal power generation exhibits both the highest capacity factor and lowest levelized cost of *any* power source. In spite of these important advantages, its potential for adding reliable baseload power has barely been tapped. The biggest cost associated with geothermal power development is usually drilling. This factor has historically restricted most development to regions where geothermal resources are located close to the surface. However, geothermal resources can actually be developed anywhere provided that the developer is willing to drill deep enough to access the thermal resource.

Hundreds of thousands of nonproducing oil and gas wells have already been drilled all over the world into geothermal reservoirs. The depths at which the drillers encountered geothermal resources in the form of steam and hot water are well documented in well logs. Some of these wells would certainly offer strong potential for geothermal power development and greatly reduce the overall cost of system construction. Additionally, new well stimulation techniques developed in the rapidly growing shale oil and gas industry could further enhance the feasibility of some of these resources (Sutra *et al.*, 2017). Chevron began dabbling in geothermal energy in the 1960s, by pioneering the development of resources associated with geysers, north of San Francisco, California. It has since developed significant geothermal projects in the Philippines and Indonesia. However, its efforts have focused primarily on leveraging its drilling capabilities to exploit resources near significant volcanic activity where geothermal resources are more easily accessed (Jay, 2016). In late December 2016, Chevron announced the sale of its geothermal assets, valued at US$3 billion, to Ayala Corp, a

Philippines-based banking conglomerate, and Star Energy, an Indonesia-based independent energy company (Dela Cruz, 2016).

Most oil and gas companies do not currently play an active role in the geothermal energy business. No system is in place to transfer depleted oil and gas wells to a geothermal energy division or partner that can develop the resource. The vast majority of these unused wells are still in the possession of the oil and gas industry. They are aware of the vast potential for this resource but are unlikely to develop it in the near future because the geothermal energy source would compete directly with natural gas. Consequently, extensive development of geothermal systems could potentially contribute to stranding vast amounts of natural gas assets. However, given the clear advantages of this opportunity, it is inevitable that a forward-thinking organization will take advantage of this opportunity in the not-too-distant future.

The future success of organizations will increasingly require a system perspective that connects end to end from the supply chain through customer utilization via a circular flow of energy and materials. System components will be integrated and optimized to generate maximum benefits, with minimum costs and negative impacts, by focusing on providing value and function to customers instead of "stuff." Organizations that continue to focus only on producing and selling stuff will suffer while organizations that evolve by embracing technology and using it to optimize the lifecycle of systems will prosper. Of course, their prosperity will be dependent on their ability to generate revenue from the function and performance of the improved systems.

In order to stay effective, competitive, and resilient, organizations will need to regularly assess the true needs of their customers and their capabilities for addressing those needs. They need to stay current with respect to emerging innovations that can more effectively optimize systems to meet customer needs. When innovations surface that can be implemented to increase effectiveness beyond current product and service offerings, organizations need to adjust their business models accordingly or they will be at considerable risk. In particular, organizations need to continuously assess how they can use the growing information resource to optimize their systems to improve effectiveness. Finally, they need to continuously assess innovation opportunities in adjacent sectors and identify partnerships that can optimize interactions between sectors and develop more sustainable systems.

**13**

# Sustainability principle #4: Restore value

> Sorrows gather around great souls as storms do around mountains; but, like them, they break the storm and purify the air of the plain beneath them.
>
> Jean Paul

The guiding principles described in the previous three chapters focus on practices that can maximize benefits while minimizing costs, resource consumption, and negative impacts. Aligning an organization's strategy with these principles is critical for ensuring long-term endurance, and can ensure that resources are conserved and protected for the future. Unfortunately, these three principles do not address the considerable degradation that has already occurred to community, environmental, and economic assets. Previous and ongoing negligence has depleted many assets to the point where slowing the rate of degradation simply prolongs catastrophe. More transformational change is needed to restore the health and value of many degraded resources. The investment required to restore the value of resources can be quite significant but, when done well, can produce multiple economic, environmental, and social dividends that are well worth the investment. Restoration initiatives and projects can range from restoring the health of an individual person to restoring the quality of life in a neighborhood to restoring lands. Regardless of the application, this principle goes beyond stopping the bad behaviors that are causing damage to adopting good behaviors that restore value.

Most of us, at one time or another, have known an individual that is morbidly obese because they consume far too much and exercise far too little. We wish they would change their ways because failure to do so will almost certainly end in tragedy. Along the way, they will experience declines in quality of life as their heart weakens, their circulatory system clogs, their blood chemistry vacillates, and their joints degrade. The problems are interconnected such that declines in one aspect create complications that lead to additional declines in other aspects. As these declines progress, they lose their ability to make valuable contributions to their families and communities. While each of the physical ailments can be slowed or mitigated to some extent through medical treatments, their health will not be restored without proactive transformational change. Incremental lifestyle improvements will only prolong the calamity that awaits them at the end of their uncomfortable and tragic journey. Transformational lifestyle changes are necessary or the damage becomes irreversible, or a catastrophe occurs.

To actually restore their health, morbidly obese individuals have to cease their destructive behavior and engage in more constructive habits and activities. There are countless inspiring examples of individuals who have been able to make the kinds of transformation needed to restore their health and improve both their longevity and quality of life. As their health is restored, their families, friends, and communities all benefit from their improved productivity and contributions, along with their reduced need for support and assistance. In spite of the sacrifice and discomfort that goes along with this sort of transformation, individuals who have successfully gone through it rarely regret doing so because the benefits of improved health far outweigh the costs.

For a morbidly obese person, excessive consumption and inactivity lead to health degradation and a progressive decline in their quality of life. The evolution of this dynamic is similar to the way that excessive consumption and wastefulness amplify constraint headwinds that decline the quality of our communities, our environment, and our economy. Just as doctors can prescribe treatments that will help obese people reduce their suffering, governments around the world have mounted massive campaigns that attempt to mitigate various aspects of constraint headwinds. Much of their focus has been on trying to improve community health and environmental quality by reducing declines in resource quality and availability.

Countless programs have been undertaken to address the endless list of social and environmental ailments that affect today's world. These include programs focused on addressing poverty, food shortages, water scarcity, lack of sanitation, homelessness, refugees, human rights, expensive and inadequate healthcare, crime, ineffective education, inequality, ecosystem destruction, pollution, climate change, and so on. While each of these individual ailments can be mitigated to some extent through government and philanthropic programs, overall the problems continue to escalate in many parts of the world. The problems are interconnected such that declines in one aspect produce complications that can lead to further declines in other aspects. Simply slowing the rate of degradation by minimizing our footprints is no longer an adequate strategy because the extent of degraded resources is simply too great.

Transformational changes are needed to reverse problematic trends and restore health and value to degraded resources. In many cases, restoration opportunities can present compelling prospects for organizations to improve the economic, social, and environmental aspects of assets. The list of assets in need of restoration is immeasurable and a comprehensive description of the need is well beyond the scope of this book. Therefore, several representative examples are provided below that describe how effective restoration initiatives have successfully restored degraded social, environmental, economic, and human assets.

## The need for land restoration

Nowhere is the need for value restoration more evident than in the world's degraded lands problem. Centuries of negligent land management practices have led to massive degradation on all continents. The UN estimates 25% of the world's land is now considered to be "highly degraded" due to poor management practices that have resulted in deforestation, desertification, wetland destruction, severe erosion, and contamination. Critical habitat has been lost, impacting countess species that count on it for survival. Urban landscapes are littered with derelict industrial sites that have either been abandoned or remain in their degraded state due to economic crises or disinterest. Over half of the globe's wetlands have been lost since 1900. Wetlands are critical resources for naturally filtering water and controlling floods by righting rivers and streams to effectively deal with storm-water run-off (FAO, 2011). Coastal wetlands serve as natural surge protectors that provide resilience against major storms and their depletion contributed greatly to disastrous impacts associated with recent hurricanes such as Sandy and Katrina.

Loss of vegetation and soil humus from degraded lands has contributed greatly to atmospheric greenhouse gas accumulation as the carbon-rich vegetation and soil humus decomposes and converts to $CO_2$. This loss of vegetation has a twofold contribution to atmospheric greenhouse gas increases through $CO_2$ release from decomposing biomass and the lack of vegetation to absorb $CO_2$ through photosynthesis (Lindsey, 2015). The potential for productively sequestering atmospheric carbon in soils and ecosystems where it is very valuable is quite compelling. According to ecologists at the Woods Hold Research Center, forests could potentially absorb 50% of annual global $CO_2$ emissions over the next 35 years. Additional amounts can be absorbed by grasslands and agricultural soils (Houghton *et al.*, 2016). The concept of sequestering carbon in natural landscapes was a major component of President Obama's 2009 Global Climate Change Initiative. However, as the administration's focus switched to the Clean Power Plan, it chose to disregard the sustainable landscape opportunity.

Reductions in economic vitality and quality of life have occurred in many parts of the world as a result of degraded lands. Restoring these resources to a state that functions ecologically – and provides the resilience, materials, and services needed by communities and businesses – would make a significant contribution to the health of our economy, our communities, and our planet. Restoration of the degraded lands will be crucial in the future, given the additional 2 billion inhabitants that will occupy the planet over the next several decades (Lindsey, 2015).

Multiple government agencies and NGOs have responded to this crisis with initiatives to reduce or stop deforestation and prevent further degradation through better land management practices. These are certainly worthwhile programs that can help slow the rate of degradation but they do not address the lands that are already degraded. In recent years, NGOs such as the Nature Conservancy and the World Resources Institute have instituted restoration initiatives. Likewise, some government agencies such as the U.S. Army Corps of Engineers have incorporated restoration into their programs. In the private sector, Dow has successfully undertaken some very innovative restoration projects.

No coherent strategy currently exists to deal with the massive need for land restoration at either national or global levels. Although successful restoration projects are now occurring all over the world, they are usually undertaken at a small scale and often in isolation from each other. In many instances, a compelling business case can be made to restore degraded properties. In the U.S.A. alone, BenDor *et al.* (2014) assessed the estimated national direct spending to be about US$10.6 billion annually.

The following case study provides an excellent example of how restoring a degraded property can produce multiple social, economic, and environmental benefits.

**Case study: Queen Elizabeth Olympic Park**

When London won the bid for the 2012 Olympic Games, its leaders made a bold move with respect to site selection by choosing a 618-acre industrial brownfield site in East London for the location of the Queen Elizabeth Olympic Park. The site had a long history of heavy industrial use, landfills, hazardous waste, and degraded waterways. It was devastated by heavy bombing during World War II and used to store demolition debris from bomb-damaged buildings. It served as a hazardous waste landfill for over 100 years, surrounded by low-income residents living in poor conditions. By integrating plans for the Queen Elizabeth

Olympic Park with long-term plans developed by the London Mayor's office, the project has become one of the most transformational developments in Europe. The park creates an oasis in the middle of one of the world's most diverse and densely populated cities. It includes five world-class sporting venues, thousands of new homes, new business districts, a world-class culture and university district, a new media and digital hub, and 111 acres of parkland (Greater London Authority, 2017).

Before construction of the infrastructure and facilities could commence, over 215 buildings had to be demolished, as well as a number of walls, bridges, and roads. About 98.5% (425,000 tons) of the demolition material, comprised mostly of steel, concrete, and masonry, was reused or recycled. Another major part of the site preparation included remediation of the contaminated soils and groundwater. The world's largest soil-washing operation was implemented on-site to decontaminate 3 million cubic yards of material for reuse (DCMS, 2012).

After much debate among the project's many stakeholders, six principal design themes were emphasized:

1. Infrastructure and urban form
2. Connectivity
3. Topography
4. Water
5. Vegetation and biodiversity
6. Use and activity

A total of 29 new bridges and underpasses were built using reclaimed materials to connect local communities. Natural features such as swales and reed beds have been designed into the landscape to perform the dual functions of promoting biodiversity and mitigating flood risk. They sit within a network of broad meadows, fields, and intricate wetlands. Two wetland areas were established and planted with 350,000 wetland and borderline aquatic plants grown from seeds and cuttings (DCMS, 2012).

In total, about US$10 billion of public and private-sector investment went into the construction and infrastructure for Queen Elizabeth Olympic Park. During its construction over 80,000 workers were utilized on the project. In addition to providing an excellent venue for the 2012 Olympics, it has provided a catalyst for additional restoration and development, and it is expected that there will be 15,000 jobs created on the park by 2025 (Mayor of London, 2017a). By 2030, the park is projected to be home to more than 10,000 new households constructed in five neighborhoods planned around green spaces and squares. They will include contemporary homes that take their cue from London's traditional Georgian and Victorian squares and terraces. Buildings will connect to an energy-efficient heat network, while utilizing efficient appliances and lighting. Electric vehicle charging stations will be provided and overall residential $CO_2$ emissions are predicted to be at least 50% lower than the 2010 Building Regulations target emission rate (Mayor of London, 2017b).

## The need for restoring self-worth

The U.S. Bureau of Justice estimates that about 1.6 million prisoners are incarcerated in the U.S.A. This represents the highest incarceration rate in the world, with about 500 prisoners per 100,000 residents compared with the rest of the developed world, which averages about 100 prisoners per 100,000 residents. Men make up

90% of the prison and local jail population. These men are overwhelmingly young (in their 20s and 30s) and less educated. The average state prisoner has a tenth-grade education and 70% have not completed high school. Incarceration rates are significantly higher for blacks and Latinos than for whites. In 2010, black men were incarcerated at a rate of 3,074 per 100,000 residents, Latinos at a rate of 1,258 per 100,000, while white men were incarcerated at a rate of 459 per 100,000 (PRB, 2012).

Incarceration is very costly to society, with the average cost for Federal inmates in fiscal year 2015 estimated at US$31,977.65, or US$87.61 per day (Federal Register, 2016). Additionally, the incarceration process has not proven to be particularly effective for prisoner rehabilitation. In a study of over 400,000 prisoners that were released from prisons in 30 states during 2005, the U.S. Bureau of Justice reported that over half were rearrested within a year, two-thirds were rearrested within three years, and three-quarters were rearrested within five years (National Institute of Justice, 2017).

Given the expense – and lack of effectiveness – regarding the incarceration system, many prevention-based programs have been developed in an effort to reduce the U.S. incarceration rate. The Office of Juvenile Justice and Delinquency Prevention recommends the following types of school and community prevention program to prevent juvenile delinquency (Youth.gov, 2017):

- Classroom and behavior management programs
- Multicomponent classroom-based programs
- Social competence promotion curriculums
- Conflict resolution and violence prevention curriculums
- Bullying prevention programs
- After-school recreation programs

- Mentoring programs
- School organization programs
- Comprehensive community interventions

According to a study by the National Institute of Justice, 229 programs have been undertaken in recent years that focus on delinquency prevention. They assessed that 50 of them have been effective (National Institute of Justice, 2016). Progress achieved by the prevention programs has been slow but steady: the Bureau of Justice estimates that the correctional population has declined by an annual average of 1% since 2007 (Kaeble *et al.*, 2015).

It is imperative that society finds more effective ways to rehabilitate incarcerated people so they can become valuable members of society. The following case study shows how a flooring manufacture partnered with the South Carolina prison system to accomplish just that.

**Case study: Manufacturing wood flooring in prisons**

Family-owned Anderson Hardwood Floors is the third-largest engineered-hardwood flooring manufacturer in the U.S.A., producing more than US$60 million of flooring each year. The third-generation, family-owned company employs 300 people at two company-owned facilities along with two factories at South Carolina prisons. Anderson started manufacturing flooring within the South Carolina prison system in 1996.

In the early 1990s, Anderson's sales had grown to the point where it needed to add an additional shift to its manufacturing operations. It had difficulty finding quality employees to fill the second shift and experienced a variety of human resource problems including absenteeism, drugs, and alcohol. Anderson worked with the South Carolina Department of Corrections to explore the possibility of utilizing South Carolina prisoners as a labor source through participation in a federal program called Prison Industries Enhancement (PIE). The PIE program is designed to put inmates to work in a real-world working environment. They are provided an opportunity to learn a trade while they repay their debt to society. They learn valuable job skills in a true production setting while also maintaining quality control. The program is entirely voluntary but the jobs are highly coveted by the inmates, with ten inmates applying for each opening offered by Anderson.

The availability and dependability of the prison labor provides Anderson flexibility in scheduling. It also allows Anderson to cost-effectively develop new products that require more intensive labor input, such as flooring planks, which are hand scraped by inmates to give them a weathered look. Attrition is rare because the prisoners tend to appreciate the job more and they also tend to do a better job. Less than a dozen people have been kicked out of the program since its inception. Safety incidents have been virtually non-existent, inmates are screened for drugs and alcohol, and they are almost never late or absent. Anderson has been so pleased with the inmates' performance that it permanently hired six of them on their release from prison. Customers have been uniformly pleased with the program and find it to be well aligned with their values.

Inmates earn the same prevailing wages as civilians in the local community and can qualify for bonuses if production targets are met. Some of the pay goes to the state for room and board, and a portion is earmarked for crime victim reparations. Prisoners are also able to send money home to their families, so the families stay engaged and are happy to see them when they are released. Additionally, savings accounts are established for each of the prisoners to help their transition back into society after their prison term is completed. Since 1996, Anderson has paid US$7.3 million in inmate wages.

The program has proven to be an effective means for changing the pattern of behavior that landed the inmates in prison. It affords the inmates an opportunity to put their time to productive purpose with benefits that extend well beyond their prison employment. The average recidivism rate for inmates that participate in this program is less than 7% compared with the national recidivism rate that exceeds 50%. Given that the average cost of incarceration in the U.S. is approaching US$32,000 per year, the savings for society are extraordinary (Miller, 2011).

The global need for restoration of degraded social, environmental, and economic assets is immeasurable. The examples described above are provided in the hope that they will inspire readers to explore restoration initiatives and projects that are relevant to their circumstances and values. The problems described in the case studies are representative of the vast magnitude and intensity of many degradation problems. The measures taken to restore value in each case were both inspiring and effective. However, it is important to recognize that value restoration opportunities can be much smaller and more personal. From mentoring troubled youth to participating in a community clean-up day, absolutely everyone can contribute to restoring the value of degraded social, environmental, and economic assets.

# 14
# A call for leadership

> Integrity is not a conditional word. It doesn't blow in the wind or change with the weather. It is your inner image of yourself.
>
> John D. MacDonald

The guiding principles described in the previous four chapters explain some very practical approaches and examples that organizations can use to embed sustainability principles within a wide range of aspects and functions. These principles are designed to be compatible across very diverse sectors and cultures. After all, what organization isn't interested in preventing waste, improving quality, optimizing systems, and restoring value? While these principles are well aligned with the values of most organizations, that doesn't mean that individuals, on reading this book, will march off and use this compass to drive change in their organizations.

Change involves risk, and accepting risk requires leadership. A leadership crisis exists throughout the world today with respect to effectively dealing with constraint headwinds. The international community has largely failed to address any major global issue in recent years. It has failed to deal effectively with global warming and the struggling global economy, while civil unrest grows and violence continues to fester in the Middle East. As governments have grown, their mechanisms have been plagued by decades of sectarian alignment, dynasty building, and deep corruption. They have focused all too often on deploying technology (e.g., communication infrastructure) that they believe will grow their economies while neglecting the infrastructure needed for basic quality of life improvements. Consequently, more people now have access to a mobile phone than to a toilet (WHO and UNICEF, 2015).

In the private sector, many organizations have systematically worked to discourage leadership by focusing on the management of current assets and future risks instead of innovation that can create breakthrough improvements. A great deal of corporate leadership has evolved to become a prescriptive, politically correct version of the real thing. Most of today's emerging leaders have learned from experience that driving innovation is not the most productive way to climb the corporate ladder. They have deduced that the path to career success is facilitated by avoiding conflict and creating the right image, or "corporate presence," for themselves. They focus on wearing the right clothes, passing out disingenuous compliments, and nodding their heads up and down when their bosses speak. They follow established procedures religiously and avoid taking risks or driving innovation. For many organizations, these tactics have proven to be the quickest way to climb the ladder of success. Unfortunately, the ladder of success that so many are driven to climb is often leaning against the wrong wall. Top management and governance boards often plead for more innovation because they recognize its importance for future competitiveness in our rapidly changing society. As the world-renowned management guru Jack Welch once explained: "If the rate of change on the outside exceeds the rate of change on the inside, the end is near."

Using sustainability principles to guide an organization's strategy represents a disruptive change for many organizations. **Many people can find problems, some can find solutions, but few have the courage to drive change.** Innovation by its very nature requires risk because it involves making changes that are frequently disruptive. From the perspective of an individual employee, the potential consequences associated with failure far outweigh any potential benefits. In cases where attempts at innovation are successful, the benefits are usually claimed by the employees' superiors whereas the discomfort that always accompanies disruptive change is assigned to the employee. In a worst-case scenario, where the attempt at innovation happens to fail, the employee who championed the innovation will own the failure ... *forever*!

Employees need a safe place to fail if we want them to take the risks that come with innovation and true leadership. Unfortunately, most of today's performance evaluation systems are structured to penalize failure and disruption while rewarding employees who don't make mistakes and stay clear of controversy. Employee performance is frequently assessed from a *teacher–student* perspective. If a student doesn't make errors, they are rewarded with a good grade. This dynamic repeats itself throughout an individual's academic experience and usually carries over into the professional world. In my opinion, a more productive dynamic is that of the *coach–player*. Good coaches know that their team's success depends on the ability of their players to make plays.

Simply striving to avoid errors will not lead to enduring success. Good coaches encourage their players to push their limits and try new techniques in training, practice, and competition. They recognize that players will frequently fail in their attempts to improve but accept these failures as part of the learning and improvement process. Excellence can only be achieved by taking chances. A great baseball or softball player will probably fail in two-thirds of their plate appearances, and many of the world's greatest basketball players miss over half of their shots. Yet their coaches wouldn't dream of telling them to stop swinging at good pitches or taking good shots. Innovation is about taking these kinds of chances – you can't score if you never take a shot. Real leaders don't possess an unhealthy fear of failure – they encourage team members to take risks. Mike Myatt (2013) summarized these dynamics well:

> Until organizations reject those playing leadership and embrace those willing to challenge the status quo, offer new thought, encourage dissenting opinion, and who desire to serve instead of seeking to be served, we'll continue to see organizations struggle unnecessarily. When process becomes more important than people, when collaboration is confused with having a meeting, when potential is held in higher regard than performance, and when independent thinking takes a backseat to conformity, leadership is dysfunctional at best. Leadership simply cannot be engineered according to the mass adoption of a set of rules (best practices). Leadership is about breaking the rules to discover change and innovation (next practices).

One of the primary reasons that leadership is deficient in today's world is the uncertainty regarding near-term economic, environmental, and social conditions. Uncertainty adds risk and most leaders are risk averse. For instance, long-term forecasts have predicted for many years that growing population and an expanding middle class will create resource shortages for many commodities. Yet the world is currently experiencing an unanticipated global glut of many commodities, including oil, natural gas, coal, and most metals. Each time these conflicting projections emerge, confidence in future trends erodes and leaders choose to stay with the status quo. They elect not to invest in promising new opportunities because they are concerned that their assumptions about future conditions will not prove to be accurate and the investment will not prove to be fruitful. However, the risks of taking strategic actions are minimized when decisions are guided by sound principles such as those described in this book. Time-tested principles based on widely accepted values do not change regardless of rapidly changing economic, environmental, and social conditions. Therefore, leadership decisions guided by sound principles will almost always be appropriate in the long term, provided the actions driven by the decisions are well planned and executed. Future risk can be minimized with this approach regardless of uncertainty regarding short-term fluctuations in economic, social, and environmental conditions.

The sustainability compass described in this book is based on sound principles that are time-tested. Using these principles to guide strategy and innovation can provide reliable guidance for making decisions that are directionally correct. Uncertainty is reduced by following these principles because leaders can be confident that they are on a course of enduring success even though the world around them is constantly changing and creating new challenges. Utilizing these principles to guide strategy and innovation is a proven means for diminishing the effects of constraint headwinds. In many cases, following these principles can place an organization on a course where the headwinds that constrain progress for others can become a productive source of prosperous tailwinds that can energize progress and ensure effectiveness.

---

May the road rise up to meet you.
May the wind always be at your back.
May the sun shine warm upon your face,
and rains fall soft upon your fields.
And until we meet again,
May God hold you in the palm of His hand.

Old Irish Blessing

---

# Bibliography


3M (2017). 3M Sustainability. Retrieved April 19, 2017 from http://www.3m.com/3M/en_US/sustainability-us/goals-progress/

Accenture Strategy (2016). The UN Global Compact: Accenture Strategy CEO Study. Retrieved April 18, 2017 from https://www.accenture.com/us-en/insight-un-global-compact-ceo-study

ACEEE (American Council for an Energy-Efficient Economy) (2017). The International Energy Efficiency Scorecard. Retrieved April 18, 2017 from http://aceee.org/portal/national-policy/international-scorecard

Alcorta, L., M. Bazilian, and G. De Simone (2014) Return on investment from industrial energy efficiency: evidence from developing countries. *Energy Efficiency*, 7(1), 43-53.

Alexandratos, N., and J. Bruinsma (2012). *World Agriculture Towards 2030/2050: The 2012 Revision*. ESA Working Paper No. 12-03. Rome, Italy: FAO.

The Aluminum Association (2017). Recycling. Retrieved April 18, 2017 from http://www.aluminum.org/industries/production/recycling

American Iron and Steel Institute (2006). *The New Steel: Sustainable, World Leader in Recycling*. Retrieved April 18, 2017 from https://www.steel.org/~/media/Files/AISI/Fact%20Sheets/fs_sustainable_recyclable_oct08.pdf

American Sustainable Business Council (2015). *Making the Business & Economic Case for Safer Chemistry*. Retrieved April 19, 2017 from http://asbcouncil.org/sites/default/files/asbcsaferchemicalsreportpresred.pdf

Amoranto, G., N. Chun, and A. Deolalikar (2010). *Who Are the Middle Class and What Values Do They Hold? Evidence from the World Values Survey*. Retrieved April 17, 2017 from https://think-asia.org/handle/11540/1566

Arabella Advisors (2016). *The Global Fossil Fuel Divestment and Clean Energy Investment Movement*. Retrieved April 18, 2017 from https://www.arabellaadvisors.com/wp-content/uploads/2016/12/Global_Divestment_Report_2016.pdf

Asian Development Bank (2010). The rise of Asia's middle class. In *Key Indicators for Asia and the Pacific* (pp. 29-36). Manila, Philippines: Asian Development Bank.

Baldé, C.P., F. Wang, R. Kuehr, and J. Huisman (2015), *The Global E-Waste Monitor – 2014*. Bonn, Germany: United Nations University, IAS – SCYCLE.

Barrett, D. (2015, October 10). U.S., BP finalize $20.8 billion Deepwater oil spill settlement. *The Wall Street Journal*. Retrieved April 18, 2017 from https://www.wsj.com/articles/u-s-says-20-8-billion-bp-spill-settlement-finalized-1444058619

BBC News (2016, April 11). Tata Steel to launch sale of UK plants. *BBC News*. Retrieved April 17, 2017 from http://www.bbc.com/news/business-36009437

Beck, E.C. (1979). The Love Canal tragedy. Retrieved April 20, 2017 from https://archive.epa.gov/epa/aboutepa/love-canal-tragedy.html

BenDor, T.K., T.W. Lester, and A. Livengood (2014). Exploring and understanding the restoration economy. Final report to Walton Family Fund. Retrieved May 8, 2017 from https://curs.unc.edu/files/2014/01/RestorationEconomy.pdf

Bertoncello, M., and D. Week (2015). Ten ways autonomous driving could redefine the automotive world. Retrieved April 19, 2017 from http://www.mckinsey.com/industries/automotive-and-assembly/our-insights/ten-ways-autonomous-driving-could-redefine-the-automotive-world

Bioplastics Magazine (2015, December 5). Bio-based production capacity to soar to 17 million metric tons in 2020. *Bioplastics Magazine*. Retrieved April 18, 2017 from http://www.bioplasticsmagazine.com/en/news/meldungen/20150512-nova-Institute-publishes-market-study-update.php

Bloomberg (2015, June 1). Even Big Oil wants a carbon tax. *Bloomberg*. Retrieved April 17, 2017 from https://www.bloomberg.com/view/articles/2015-06-01/even-big-oil-wants-a-carbon-tax

Borick, C., B.G. Rabe, and S.B. Mills (2015). Acceptance of global warming among Americans reaches highest level since 2008: A report from the National Surveys on Energy and Environment. *Issues in Energy and Environmental Policy*, 25, 1-8.

Bouton, S., S.M. Knupfer, I. Mihov, and S. Swartz (2015). Urban mobility at a tipping point. Retrieved May 8, 2017 from http://www.mckinsey.com/business-functions/sustainability-and-resource-productivity/our-insights/urban-mobility-at-a-tipping-point

Budischak, C.D., H. Sewell, H. Thomson, L. Mach, D.E. Veron, and W. Kempton (2013). Cost-minimized combinations of wind power, solar power and electrochemical storage, powering the grid up to 99.9% of the time. *Journal of Power Sources*, 225(1), 60-74.

Bulbs.com (2017). LED FAQs. Retrieved April 19, 2017 from http://www.bulbs.com/learning/ledfaq.aspx

Bullitt County History (2017). Valley of the Drums. Retrieved April 17, 2017 from http://bullittcountyhistory.org/bchistory/valleydrum.html

Bureau of International Recycling (2017). The industry. Retrieved April 18, 2017 from http://www.bir.org/industry/

Byrd, H. (2007). *A Comparison of Three Well Known Behavior Based Safety Programs: DuPont STOP Program, Safety Performance Solutions and Behavioral Science Technology* (Unpublished thesis). Rochester Institute of Technology, Rochester, NY. Retrieved April 17, 2017 from http://scholarworks.rit.edu/theses/686

Caterpillar Inc. (2017). Making sustainable progress possible. Retrieved April 18, 2017 from http://www.cat.com/en_US/company.html

CDP (Carbon Disclosure Project) (2016). *CDP Global Climate Change Report 2015: At the Tipping Point?* Retrieved April 18, 2017 from https://www.cdp.net/en/research/global-reports/global-climate-change-report-2015

Chamberlin, A. (2014, September 10). BP lost 55% shareholder value after the Deepwater Horizon incident. *Market Realist*. Retrieved April 18, 2017 from http://marketrealist.com/2014/09/bp-lost-55-shareholder-value-deepwater-horizon-incident/

Chemical Strategies Partnership (2017). About CSP. Retrieved April 19, 2017 from http://www.chemicalstrategies.org/about.php

Citi (2015). *Energy Darwinism II: Why a Low Carbon Future Doesn't Have to Cost the Earth*. New York, NY: Citi Global Perspectives and Solutions.

Construction and Demolition Recycling Association (2014). Concrete recycling. Retrieved April 18, 2017 from http://www.cdrecycling.org/concrete-recycling

The Conversation (2012, September 9). Saving the ozone layer: why the Montreal Protocol worked. Retrieved April 17, 2017 from http://theconversation.com/saving-the-ozone-layer-why-the-montreal-protocol-worked-9249

Covey, S.R. (2014). *The Seven Habits of Highly Effective People: Restoring the Character Ethic*. West Valley City, UT: FranklinCovey.

CPSC (U.S. Consumer Products Safety Commission) (2017). About CPSC. Retrieved April 17, 2017 from https://www.cpsc.gov/about-cpsc

Cramer, J., and A. Krueger (2015). *Disruptive Change in the Taxi Business: The Case of Uber*. IRS Working Papers 595. Retrieved April 19, 2017 from http://arks.princeton.edu/ark:/88435/dsp01v692t860d

DCMS (Department for Culture, Media and Sport) (2012). *Beyond 2012: The London 2012 Legacy Story*. Retrieved April 20, 2017 from https://www.gov.

uk/government/uploads/system/uploads/attachment_data/file/77993/DCMS_Beyond_2012_Legacy_Story.pdf

de Groot, R., L. Brander, S. van der Ploeg, R. Costanza, F. Bernard, L. Braat, M. Christie, N. Crossman, A. Ghermandi, L. Hein, S. Hussain, P. Kumar, A. McVittie, R. Portela, L. Rodriguez, P. ten Brink, and P. van Beukering (2012). Global estimates of the value of ecosystems and their services in monetary units. *Ecosystem Services*, 1(1), 50-61.

Dela Cruz, E. (2016, December 23). Indonesia, Philippine groups acquire Chevron's $3 billion geothermal assets. *Reuters*. Retrieved May 8, 2017 from http://www.reuters.com/article/us-chevron-sale-geothermal-idUSKBN14C0OW

Deming, W.E. (1982). *Out of the Crisis*. Cambridge, MA: MIT Press.

Dobbs, R., S. Lund, J. Woetzel, and M. Mutafchieva (2015, February). Debt and (not much) deleveraging. Retrieved April 17, 2017 from http://www.mckinsey.com/global-themes/employment-and-growth/debt-and-not-much-deleveraging

DuPont (2017). DuPont™ STOP®. Retrieved April 19, 2017 from http://www.training.dupont.com/dupont-stop

The Economist (2015a, April 22). Recycling in America. *The Economist*. Retrieved April 18, 2017 from http://www.economist.com/blogs/democracyinamerica/2015/04/recycling-america

——— (2015b, July 1). If autonomous vehicles rule the world: from horseless to driverless. *The Economist*. Retrieved April 19, 2017 from http://worldif.economist.com/article/12123/horseless-driverless

Edison Energy (2016, May 5). Energy-as-a-service: the new paradigm shift. Retrieved April 19, 2017 from http://www.edisonenergy.com/blog/energy-service-new-paradigm-shift/

EIA (U.S. Energy Information Administration) (2015). What is U.S. electricity generation by energy source? Retrieved April 18, 2017 from https://www.eia.gov/tools/faqs/faq.php?id=427&t=3

——— (2016a). *International Energy Outlook 2016*. Washington, DC: EIA.

——— (2016b, April 28). Power sector coal demand has fallen in nearly every state since 2007. Retrieved April 17, 2017 from https://www.eia.gov/todayinenergy/detail.php?id=26012

——— (2016c). *Levelized Cost and Levelized Avoided Cost of New Generation Resources in the Annual Energy Outlook 2016*. Retrieved April 18, 2017 from https://www.eia.gov/outlooks/aeo/pdf/electricity_generation.pdf

——— (2017). How much carbon dioxide is produced when different fuels are burned? Retrieved April 17, 2017 from https://www.eia.gov/tools/faqs/faq.php?id=73&t=11

EPA (U.S. Environmental Protection Agency) (1983, February 22). *Joint Federal/State Action Taken to Relocate Times Beach Residents* [press release]. Retrieved

April 17, 2017 from https://archive.epa.gov/epa/aboutepa/1983-press-release-joint-federalstate-action-taken-relocate-times-beach-residents.html

——— (2007). Frequent questions about EPA's risk assessment of spent foundry sands in soil related applications. Retrieved April 18, 2017 from https://www.epa.gov/smm/frequent-questions-about-epas-risk-assessment-spent-foundry-sands-soil-related-applications

——— (2016). Superfund history. Retrieved April 17, 2017 from https://www.epa.gov/superfund/superfund-history

——— (2017a). Ozone layer protection. Retrieved April 17, 2017 from https://www.epa.gov/ozone-layer-protection

——— (2017b). Sustainable management of construction and demolition materials. Retrieved April 18, 2017 from https://www.epa.gov/smm/sustainable-management-construction-and-demolition-materials

——— (2017c). CHP benefits. Retrieved April 18, 2017 from https://www.epa.gov/chp/chp-benefits

——— (2017d). Causes of climate change. Retrieved April 18, 2017 from https://www.epa.gov/climate-change-science/causes-climate-change#changessun

——— (2017e). WasteWise Program. Retrieved April 19, 2017 from https://archive.epa.gov/epawaste/conserve/smm/wastewise/web/pdf/insidepages.pdf

European Commission (2016). Non-financial reporting. Retrieved April 18, 2017 from http://ec.europa.eu/finance/company-reporting/non-financial_reporting/index_en.htm

FAO (Food and Agriculture Organization of the United Nations) (2011). *The State of the World's Land and Water Resources for Food and Agriculture: Managing Systems at Risk*. London: Earthscan.

Federal Register (2016, July 19). Annual determination of average cost of incarceration. Retrieved April 20, 2017 from https://www.federalregister.gov/documents/2016/07/19/2016-17040/annual-determination-of-average-cost-of-incarceration

Fields, M. (2016, August 16). Ford's road to full autonomy. *LinkedIn*. Retrieved April 19, 2017 from https://www.linkedin.com/pulse/fords-road-full-autonomy-mark-fields

Fleming, T., and M. Zils (2014, February). Toward a circular economy: Philips CEO Frans van Houten. *McKinsey Quarterly*. Retrieved April 19, 2017 from http://www.mckinsey.com/business-functions/sustainability-and-resource-productivity/our-insights/toward-a-circular-economy-philips-ceo-frans-van-houten

ForbesBrandVoice (2014, January 7). Zipcar and car-sharing services: better than rentals? *Forbes*. Retrieved April 19, 2017 from https://www.forbes.com/sites/

northwesternmutual/2014/01/07/zipcar-and-car-sharing-services-better-than-rentals/

Franklin, B. (1999). *Wit and Wisdom from Poor Richard's Almanack*. Mineola, NY: Dover.

Gertner, J. (2014, February 10). How Philips altered the future of light. *Fast Company*. Retrieved April 19, 2017 from https://www.fastcompany.com/3025604/philips-lighting-the-way

Gray, W.B., and J.M. Mendeloff (2005). The declining effects of OSHA inspections on manufacturing injuries: 1979–1998. *Industrial and Labor Relations Review*, 58(4), 571-587.

Greater London Authority (2017). Queen Elizabeth Olympic Park: Smart London case study. Retrieved April 20, 2017 from https://www.london.gov.uk/what-we-do/business-and-economy/science-and-technology/smart-london/queen-elizabeth-olympic-park

GRI (Global Reporting Initiative) (2016a). *GRI's Contribution to Sustainable Development*. Amsterdam: Global Reporting Initiative.

——— (2016b). GRI Standards. Retrieved April 18, 2017 from https://www.globalreporting.org/standards

Gusovsky, D. (2016, March 24) America's water crisis goes beyond Flint, Michigan. *CNBC*. Retrieved April 17, 2017 from http://www.cnbc.com/2016/03/24/americas-water-crisis-goes-beyond-flint-michigan.html

Hadi, M. (2015, September 22). CARMAGEDDON: Volkswagen's unraveling has wiped billions off auto industry shares. *Business Insider*. Retrieved April 18, 2017 from http://www.businessinsider.com/volkswagen-causes-widespread-auto-drop-2015-9

Hatch, G.P. (2012). Dynamics in the global market for rare earths. *Elements*, 8(5), 341-346.

Henderson, R. (2015, June 8). What gets measured gets done. Or does it? *Forbes*. Retrieved April 18, 2017 from https://www.forbes.com/sites/ellevate/2015/06/08/what-gets-measured-gets-done-or-does-it/

Houghton, R., P. Duffy, and A. Nassikas (2016). *Forests: The Bridge to a Fossil-Free Future*. Retrieved April 20, 2017 from http://whrc.org/wp-content/uploads/2016/11/PB_Forests.pdf

Huisman, J. (2012). Eco-efficiency evaluation of WEEE take-back systems. In V. Goodship and A. Stevels (Eds.), *Waste Electrical and Electronic Equipment (WEEE) Handbook* (pp. 93-119). Cambridge, MA: Woodhead.

ILO (International Labour Organization) (2013, April 26). *ILO Calls for Urgent Global Action to Fight Occupational Diseases* [press release]. Retrieved April 17, 2017 from http://www.ilo.org/global/about-the-ilo/newsroom/news/WCMS_211627/lang--en/index.htm

Institute of Scrap Recycling Industries (2015). *Facts and Figures Fact Sheet: Recycling*. Retrieved April 18, 2017 from http://www.isri.org/docs/default-source/recycling-industry/facts-and-figures-fact-sheet---recycling.pdf?sfvrsn=16

——— (2017). Recycling industry. Retrieved April 18, 2017 from http://www.isri.org/recycling-industry

International Energy Agency (2008). *Combined Heat and Power: Evaluating the Benefits of Greater Global Investment*. Paris, France: International Energy Agency.

——— (2014). *Special Report: World Energy Investment Outlook*. Paris, France: International Energy Agency.

——— (2016a). *Key World Energy Statistics*. Paris, France: International Energy Agency.

——— (2016b). Modern energy for all. Retrieved April 17, 2017 from http://www.worldenergyoutlook.org/resources/energydevelopment/

——— (2016c, July 28). Renewable energy continuing to increase market share. Retrieved April 18, 2017 from https://www.iea.org/newsroom/news/2016/july/renewable-energy-continuing-to-increase-market-share.html

International Finance Corporation (2014). *The Business Case for Sustainability*. Retrieved April 18, 2017 from https://www.cbd.int/financial/mainstream/ifc-businesscase.pdf

ISSA (International Social Security Association) (2017). Towards a global culture of prevention. Retrieved April 17, 2017 from https://www.issa.int/en_GB/topics/occupational-risks/introduction

Jay, Y. (2016, April 22). This potential asset sale reflects Chevron's pragmatic approach to renewables. *The Motley Fool*. Retrieved May 8, 2017 from https://www.fool.com/investing/general/2016/04/22/this-potential-asset-sale-shows-chevrons-pragmatic.aspx

Kaeble, D., L.E. Glaze, A. Tsoutis, and T.D. Minton (2015). Correctional populations in the United States, 2014. Retrieved April 20, 2017 from https://www.bjs.gov/index.cfm?ty=pbdetail&iid=5519

Kesler, S. (2015, November 3). RelayRides takes a page from AirBnB, rebrands as Turo. *Fast Company*. Retrieved April 19, 2017 from https://www.fastcompany.com/3052940/relayrides-takes-a-page-from-airbnb-rebands-to-turo

Kharas, H. (2010). *The Emerging Middle Class in Developing Countries*. OECD Development Centre Working Paper No. 285. Retrieved April 20, 2017 from https://www.oecd.org/dev/44457738.pdf.

Korosec, K. (2016, January 12). Would you lease a car with 6 friends? Ford rolls out new car plan. *Fortune*. Retrieved April 19, 2017 from http://fortune.com/2016/01/12/ford-leasing-pilot/

Kuehr, R., and E. Williams (2003). *Computers and the Environment: Understanding and Managing their Impacts*. Norwell, MA: Kluwer Academic Publishers.

Kwatra, S., and C. Essig (2014). *The Promise and the Potential of Comprehensive Commercial Building Retrofit Programs*. Research Report A1402. Washington, DC: American Council for an Energy Efficient Economy.

Lindsey, T.C. (2000). Key factors for promoting P2 technology adoption. *P2: Pollution Prevention Review*, 10(1), 1-12.

——— (2007). Metal finishing and electroplating. In M. Kutz (Ed.) *Environmentally Conscious Manufacturing* (pp. 123-144). Hoboken, NJ: John Wiley & Sons.

——— (2015). *Restoring Natural Infrastructure: Strategies for Thriving Communities, Businesses and Ecosystems. A Report on Participant Observations and Recommendations from the Restoring Natural Infrastructure Summit* (held in New York on November 4, 2015). Peoria, IL: Caterpillar Inc.

MarketWatch (2016, July 12). World bioplastics market growing at 29.3% CAGR to 2020. *MarketWatch*. Retrieved April 18, 2017 from http://www.marketwatch.com/story/world-bioplastics-market-growing-at-293-cagr-to-2020-2016-07-11-2220311

Mars, C., C. Nafe, and J. Linnell (2016). *The Electronics Recycling Landscape*. Retrieved April 20, 2017 from https://www.sustainabilityconsortium.org/wp-content/uploads/2017/03/TSC_Electronics_Recycling_Landscape_Report-1.pdf.

Mayor of London (2017a). Queen Elizabeth Olympic Park: facts and figures. Retrieved April 20, 2017 from http://www.queenelizabetholympicpark.co.uk/media/facts-and-figures

——— (2017b). Queen Elizabeth Olympic Park: homes and living. Retrieved April 20, 2017 from http://www.queenelizabetholympicpark.co.uk/the-park/homes-and-living

McDonough, W., and M. Braungart (2002). *Cradle to Cradle: Remaking the Way We Make Things*. New York, NY: North Point Press.

McGregor, M., B. Brown, and M Glöss (2015). Disrupting the cab: Uber, ridesharing and the taxi industry. *Journal of Peer Production*. Retrieved May 8, 2017 from http://peerproduction.net/issues/issue-6-disruption-and-the-law/essays/disrupting-the-cab-uber-ridesharing-and-the-taxi-industry/

McKinsey (2009). *Unlocking Energy Efficiency in the U.S. Economy*. New York, NY. McKinsey.

——— (2014). Sustainability's strategic worth: McKinsey Global Survey results. Retrieved April 18, 2017 from http://www.mckinsey.com/business-functions/sustainability-and-resource-productivity/our-insights/sustainabilitys-strategic-worth-mckinsey-global-survey-results

McLaughlin, P., and R. Greene (2014, May 8). The unintended consequences of federal regulatory accumulation. Retrieved April 17, 2017 from https://www.mercatus.org/publication/unintended-consequences-federal-regulatory-accumulation

Miller, H. (2011, August 14). Anderson Hardwood Floors: inmates build new lives from the floor up. *Woodworking Network*. Retrieved April 20, 2017 from http://www.woodworkingnetwork.com/articles/anderson_hardwood_floors_-_inmates_build_new_lives_from_the_floor_up_127690638.html

Moss, R.L., E. Tzimas, H. Kara, P. Willis, and J. Kooroshy (2011). *Critical Metals in Strategic Energy Technologies: Assessing Rare Metals as Supply-Chain Bottlenecks in Low-Carbon Energy Technologies*. Luxembourg: European Union.

MSHA (U.S. Mine Safety and Health Administration) (2017a). About. Retrieved April 17, 2017 from https://www.msha.gov/about

——— (2017b). Mine inspections. Retrieved April 17, 2017 from https://www.msha.gov/compliance-enforcement/mine-inspections

Myatt, M. (2013, March 21). Why your organization suffers from leadership dysfunction. *Forbes*. Retrieved April 20, 2017 from https://www.forbes.com/sites/mikemyatt/2013/03/21/why-your-organization-suffers-from-leadership-dysfunction/

NAESCO (National Association of Energy Service Companies) (2017). What is an ESCO? Retrieved April 19, 2017 from http://www.naesco.org/what-is-an-esco

National Asphalt Pavement Association (2010). Recycling. Retrieved April 18, 2017 from http://www.asphaltpavement.org/index.php?option=com_content&task=view&id=25&Itemid=45

National Institute of Justice (2016). Juveniles: delinquency prevention. Retrieved April 20, 2017 from https://www.crimesolutions.gov/TopicDetails.aspx?ID=62

——— (2017). Recidivism. Retrieved April 20, 2017 from https://www.nij.gov/topics/corrections/recidivism/pages/welcome.aspx

National Institute of Safety and Health (2017). Prevention through design. Retrieved April 17, 2017 from https://www.cdc.gov/niosh/topics/ptd/default.html

National Intelligence Council (2013). *Global Trends 2030: Alternative Worlds*. Retrieved April 17, 2017 from https://www.dni.gov/files/documents/GlobalTrends_2030.pdf

Naughton, K. (2013, January 2). Avis Budget embraces car sharing with Zipcar acquisition. *Bloomberg*. Retrieved April 19, 2017 from https://www.bloomberg.com/news/articles/2013-01-02/avis-budget-makes-491-million-offer-to-acquire-zipcar

OECD (Organisation for Economic Co-operation and Development) (2015). *Material Resources, Productivity and the Environment*. Paris, France: OECD Publishing.

OECD.Stat (2015). Municipal waste generation and treatment. Retrieved April 18, 2017 from http://stats.oecd.org

Ohio History Central (2017). Cuyahoga River Fire. Retrieved from http://www.ohiohistorycentral.org/w/Cuyahoga_River_Fire

OSHA (U.S. Occupational Safety and Health Administration) (2001). OSHA at 30: three decades of progress in occupational safety and health. Retrieved April 17, 2017 from https://www.osha.gov/as/opa/osha-at-30.html

——— (2017). Timeline of OSHA's 40 year history. Retrieved April 17, 2017 from https://www.osha.gov/osha40/timeline.html

OSMRE (U.S. Office of Surface Mining Reclamation and Enforcement) (2017). Laws, regulations, and guidance. Retrieved April 17, 2017 from http://www.osmre.gov/lrg.shtm

PRB (Population Reference Bureau) (2012). U.S. has world's highest incarceration rate. Retrieved April 20, 2017 from http://www.prb.org/Publications/Articles/2012/us-incarceration.aspx

Procopiou, C. (2016, February 12). Air pollution claims 5.5 million lives a year, making it the fourth-leading cause of death worldwide. *Newsweek*. Retrieved April 17, 2017 from http://www.newsweek.com/55-million-deaths-air-pollution-worldwide-each-year-426159

Research and Markets (2015). *Renewable Chemicals Market: Alcohols (Ethanol, Methanol), Biopolymers (Starch Blends, Regenerated Cellulose, PBS, Bio-PET, PLA, PHA, Bio-PE, and Others), Platform Chemicals & Others – Global Trends & Forecast to 2020*. Magarpatta, India: MarketsandMarkets Research.

Risen, T. (2016, April 22). America's toxic electronic waste trade. *U.S. News and World Report*. Retrieved April 18, 2017 from https://www.usnews.com/news/articles/2016-04-22/the-rising-cost-of-recycling-not-exporting-electronic-waste

RobotEnomics (2014, February 26). Morgan Stanley: the economic benefits of driverless cars. *RobotEnomics*. Retrieved April 19, 2017 from https://robotenomics.com/2014/02/26/morgan-stanley-the-economic-benefits-of-driverless-cars/

Rockoff, H. (2000). *Getting in the Scrap: The Salvage Drives of World War II*. Department of Economics Working Paper. Newark, NJ: Rutgers University.

Rogers, E.M. (2003). *Diffusion of Innovations* (5th ed.). New York, NY: The Free Press.

Sadowsky, M. (2014). The 2013 Ratings Survey: polling the experts – A GlobeScan/SustainAbility survey. Retrieved April 18, 2017 from http://sustainability.com/our-work/reports/the-2013-ratings-survey-polling-the-experts/

SASB (Sustainable Accounting Standards Board) (2016). Home page. Retrieved April 18, 2017 from https://www.sasb.org

Science Daily (2015, February 20). Hydropower completes greening of Norway. *Science Daily*. Retrieved April 18, 2017 from https://www.sciencedaily.com/releases/2015/02/150220083914.htm

Shipley, A., A. Hampson, B. Hedman, P. Garland, and P. Bautista (2008). *Combined Heat and Power: Effective Solutions for a Sustainable Future*. Oak Ridge, TN: Oak Ridge National Laboratory.

Singer, T. (2014). *Sustainability Matters 2014: How Sustainability can Enhance Corporate Reputation*. New York, NY: The Conference Board.

——— (2015). *Driving Revenue Growth Through Sustainable Products and Service*. New York, NY: The Conference Board.

Singhal, N. (2015, September 27). As good as new. *Business Today In*. Retrieved April 18, 2017 from http://www.businesstoday.in/magazine/technology/refurbished-gadgets-as-good-as-new-market-sale-value-discount/story/223489.html

Society for Human Resource Management (2014). *Employee Job Satisfaction and Engagement: The Road to Economic Recovery*. Alexandria, VA: Society for Human Resource Management.

Stewart, J. (2016, June 15). Inside China's plan to beat America to the self driving car. *Wired*. Retrieved April 19, 2017 from https://www.wired.com/2016/06/chinas-plan-first-country-self-driving-cars/

Sustainable Biomaterials Collaborative (2017). Biobased materials and sustainability. Retrieved April 18, 2017 from http://www.sustainablebiomaterials.org/faqs.biobased.php

Sutra, E., M. Spada, and P. Burgherr (2017). Chemicals usage in stimulation processes for shale gas and deep geothermal systems: a comprehensive review and comparison. *Renewable and Sustainable Energy Reviews*, 77, 1–11.

Tannert, C. (2014, January 8). 10 autonomous driving companies to watch. *Fast Company*. Retrieved April 19, 2017 from https://www.fastcompany.com/3024362/innovation-agents/10-autonomous-driving-companies-to-watch

Turner, M. (2016, February 2) Here is the letter the world's largest investor, BlackRock CEO Larry Fink, just sent to CEOs everywhere. *Business Insider*. Retrieved May 9, 2017 from http://uk.businessinsider.com/blackrock-ceo-larry-fink-letter-to-sp-500-ceos-2016-2?r=US&IR=T

Uber (2017). Find a city. Retrieved April 19, 2017 from https://www.uber.com/cities/

UN Department of Economic and Social Affairs, Population Division (2015). *World Population Prospects, 2015 Revision*. Retrieved April 17, 2017 from https://esa.un.org/unpd/wpp/

UN Development Programme (2016). Sustainable Development Goals. Retrieved April 18, 2017 from http://www.undp.org/content/undp/en/home/sustainable-development-goals.html

UN Population Fund (2007). *State of World Population 2007: Unleashing the Potential of Urban Growth*. New York, NY: UN Population Fund.

UNEP (United Nations Environment Programme) (2011). *Decoupling Natural Resource Use and Environmental Impacts from Economic Growth*. Paris, France: UNEP.

——— (2013). *Green Economy and Trade: Trends, Challenges and Opportunities*. http://www.unep.org/greeneconomy/sites/unep.org.greeneconomy/files/field/image/fullreport.pdf

——— (2014). Second meeting of the Open-Ended Working Group. Retrieved April 19, 2017 from http://old.saicm.org/index.php?option=com_content&view=article&id=509:meeting-documents-2nd-meeting-of-the-open-ended-working-group-geneva-15-17-december-2014&catid=92:oewg

Unruh, G., D. Kiron, N. Kruschwitz, M. Reebes, H. Rubel, and A. Meyer Zum Felde (2016, May 11). Investing for a sustainable future: investors care more about sustainability than many executives believe. *MIT Sloan Management Review*. Retrieved April 18, 2017 from http://sloanreview.mit.edu/projects/investing-for-a-sustainable-future/

U.S. Environmental Protection Agency (2017). CHP benefits. Retrieved May 8, 2017 from https://www.epa.gov/chp/chp-benefits

U.S. Securities and Exchange Commission (2014). E.I. du Pont de Nemours and Company Form 10-K. Retrieved April 19, 2017 from https://www.sec.gov/Archives/edgar/data/30554/000003055415000004/dd-12312014x10k.htm#s4C4157C8843F21D742383BE22068A719

——— (2015). The Dow Chemical Company annual report on Form 10-K for the fiscal year ended December 31, 2015. Retrieved April 19, 2017 from https://www.sec.gov/Archives/edgar/data/29915/000002991516000066/dow201510k.htm#s5B25F758C4575E6D9D856CAC3ED3C9DA

USITC (U.S. International Trade Commission) (2012) *Remanufactured Goods: An Overview of the U.S. and Global Industries, Markets, and Trade*. Retrieved May 9, 2017 from https://www.usitc.gov/publications/332/pub4356.pdf

Walker, D. (2010, Fall). Understanding how tires are used in asphalt. *Asphalt*. Retrieved April 18, 2017 from http://asphaltmagazine.com/understanding-how-tires-are-used-in-asphalt/

Walmart (2017). Sustainable chemistry policy. Retrieved from http://www.walmartsustainabilityhub.com/sustainable-chemistry/sustainable-chemistry-policy

Weinschenk, C. (2016, August 12). LEDs are good. Smart lights are better. *Energy Manager Today*. Retrieved April 19, 2017 from https://www.energymanagertoday.com/leds-are-good-smart-lighting-is-better-0126247/

WHO (World Health Organization) and UNICEF (2015). *Progress on Drinking Water and Sanitation, 2015 Update and MDG Assessment*. Geneva, Switzerland: WHO.

Wilson, J., and M. Townsend (2016, February 21). Lumber Liquidators plunges after CDC raises flooring cancer risk. *Bloomberg*. Retrieved April 18, 2017 from https://www.bloomberg.com/news/articles/2016-02-21/lumber-liquidators-flooring-has-increased-cancer-risk-cdc-says

Winston, A.S. (2014). *The Big Pivot: Radically Practical Strategies for a Hotter, Scarcer, and More Open World*. Boston, MA: Harvard Business Review Press.

Waste Management (2017). Waste Management's sustainable services at-a-glance. Retrieved May 8, 2017 from http://sustainability.wm.com/waste/reduction/reduction-solutions.php#info1.

Womack, J.P., and D.T. Jones (2003). *Lean Thinking: Banish Waste and Create Wealth in Your Corporation*. New York, NY: The Free Press.

World Commission on Environment and Development (1987). *Our Common Future*. Oxford, UK: Oxford University Press.

WWF (World Wide Fund for Nature) (2016). *Living Planet Report 2016: Risk and Resilience in a New Era*. Gland, Switzerland: WWF International.

XPrize (2017). Reimagine $CO_2$: inspiring the brightest minds around the world to help solve climate change. Retrieved April 17, 2017 from http://carbon.xprize.org

Youth.gov (2017). Prevention and early intervention. Retrieved April 20, 2017 from http://youth.gov/youth-topics/juvenile-justice/prevention-and-early-intervention

Zimring, C.A. (2005). *Cash for Your Trash: Scrap Recycling in America*. New Brunswick, NJ: Rutgers University Press.

# About the author

Dr. Tim Lindsey is an internationally recognized expert regarding the use of sustainability principles and strategies to drive innovation and improve competitive advantage. Since 1980, he has helped hundreds of organizations identify and implement more sustainable innovations throughout their value chains. He specializes in helping individual business units embed sustainability principles into both corporate roles (research, product development, human resources, strategy, and risk) and operational functions (supply chain, manufacturing, quality, logistics, dealerships, and customer processes). His vast experience encompasses multiple industrial sectors, including manufacturing, mining, maintenance, energy, agriculture, food processing, biofuels, chemical processing, and electronics.

Dr. Lindsey began his career in the early 1980s, working for an engineering/construction firm that specialized in the cleanup of some of the world's most contaminated hazardous waste sites. These experiences had a profound impact on his perspective and guided him to focus his efforts upstream, on measures that can prevent such catastrophes from occurring in the first place. Ever since, he has been working to develop and deploy innovations that simultaneously improve conditions for people, communities, businesses, and the environment. He served as Caterpillar's Global Director of Sustainability from 2012 to 2016 after spending 20 years at the University of Illinois where he helped lead its sustainability research, education, and outreach efforts. Prior to his academic career, he was with Exxon for seven years at one of the world's largest mining operations, serving in the roles of Environmental Manager, Safety Manager, and Senior Project Manager. In January 2016, Dr. Lindsey founded Highlander Innovation Inc., where he and his colleagues help businesses drive and implement more sustainable innovation.